RESIDUE REVIEWS

VOLUME 88

RESIDUE REVIEWS

Residues of Pesticides and Other
Contaminants in the Total Environment

Editor
FRANCIS A. GUNTHER

Assistant Editor
JANE DAVIES GUNTHER

Riverside, California

VOLUME 88

SPRINGER-VERLAG
NEW YORK HEIDELBERG BERLIN

1983

Library of Congress Catalog Card Number 62-18595.
Printed in the United States of America.

New York: 175 Fifth Avenue, New York, N.Y. 10010
Heidelberg: 6900 Heidelberg 1, Postfach 105 280, West Germany

---

ISBN 0-387-90851-X Springer-Verlag New York Heidelberg Berlin
ISBN 3-540-90851-X Springer-Verlag Berlin Heidelberg New York

Foreword

Worldwide concern in scientific, industrial, and governmental communities over traces of toxic chemicals in foodstuffs and in both abiotic and biotic environments has justified the present triumvirate of specialized publications in this field: comprehensive reviews, rapidly published progress reports, and archival documentations. These three publications are integrated and scheduled to provide in international communication the coherency essential for nonduplicative and current progress in a field as dynamic and complex as environmental contamination and toxicology. Until now there has been no journal or other publication series reserved exclusively for the diversified literature on "toxic" chemicals in our foods, our feeds, our geographical surroundings, our domestic animals, our wildlife, and ourselves. Around the world immense efforts and many talents have been mobilized to technical and other evaluations of natures, locales, magnitudes, fates, and toxicology of the persisting residues of these chemicals loosed upon the world. Among the sequelae of this broad new emphasis has been an inescapable need for an articulated set of authoritative publications where one could expect to find the latest important world literature produced by this emerging area of science together with documentation of pertinent ancillary legislation.

The research director and the legislative or administrative advisor do not have the time even to scan the large number of technical publications that might contain articles important to current responsibility; these individuals need the background provided by detailed reviews plus an assured awareness of newly developing information, all with minimum time for literature searching. Similarly, the scientist assigned or attracted to a new problem has the requirements of gleaning all literature pertinent to his task, publishing quickly new developments or important new experimental details to inform others of findings that might alter their own efforts, and eventually publishing all his supporting data and conclusions for archival purposes.

The end result of this concern over these chores and responsibilities and with uniform, encompassing, and timely publication outlets in the field of environmental contamination and toxicology is the Springer-Verlag (Heidelberg and New York) triumvirate:

Residue Reviews (vol. 1 in 1962) for basically detailed review articles concerned with any aspects of residues of pesticides and other chemical contaminants in the total environment, including toxicological considerations and consequences.

Bulletin of Environmental Contamination and Toxicology (vol. 1 in 1966) for rapid publication of short reports of significant advances and discoveries in the fields of air, soil, water, and food contamination and pollution as well as methodology and other disciplines concerned with the introduction, presence, and effects of toxicants in the total environment.

Archives of Environmental Contamination and Toxicology (vol. 1 in 1973) for important complete articles emphasizing and describing original experimental or theoretical research work pertaining to the scientific aspects of chemical contaminants in the environment.

Manuscripts for *Residue Reviews* and the *Archives* are in identical formats and are subject to review, by workers in the field, for adequacy and value; manuscripts for the *Bulletin* are not reviewed and are published by photo-offset to provide the latest results without delay. The individual editors of these three publications comprise the Joint Coordinating Board of Editors with referral within the Board of manuscripts submitted to one publication but deemed by major emphasis or length more suitable for one of the others.

Coordinating Board of Editors

Preface

That residues of pesticide and other contaminants in the total environment are of concern to everyone everywhere is attested by the reception accorded previous volumes of "Residue Reviews" and by the gratifying enthusiasm, sincerity, and efforts shown by all the individuals from whom manuscripts have been solicited. Despite much propaganda to the contrary, there can never be any serious question that pest-control chemicals and food-additive chemicals are essential to adequate food production, manufacture, marketing, and storage, yet without continuing surveillance and intelligent control some of those that persist in our foodstuffs could at times conceivably endanger the public health. Ensuring safety-in-use of these many chemicals is a dynamic challenge, for established ones are continually being displaced by newly developed ones more acceptable to food technologists, pharmacologists, toxicologists, and changing pest-control requirements in progressive food-producing economies.

These matters are of genuine concern to increasing numbers of governmental agencies and legislative bodies around the world, for some of these chemicals have resulted in a few mishaps from improper use. Adequate safety-in-use evaluations of any of these chemicals persisting into our foodstuffs are not simple matters, and they incorporate the considered judgments of many individuals highly trained in a variety of complex biological, chemical, food technological, medical, pharmacological, and toxicological disciplines.

It is hoped that "Residue Reviews" will continue to serve as an integrating factor both in focusing attention upon those many residue matters requiring further attention and in collating for variously trained readers present knowledge in specific important areas of residue and related endeavors involved with other chemical contaminants in the total environment. The contents of this and previous volumes of "Residue Reviews" illustrate these objectives. Since manuscripts are published in the order in which they are received in final form, it may seem that some important aspects of residue analytical chemistry, biochemistry, human and animal medicine, legislation, pharmacology, physiology, regulation, and toxicology are being neglected; to the contrary, these apparent omissions are recognized, and some pertinent manuscripts are in preparation. However, the field is so large and the interests in it are so varied that the editors and the Advisory Board earnestly solicit suggestions of topics and authors to help make this international book-series even more useful and informative.

"Residue Reviews" attempts to provide concise, critical reviews of timely advances, philosophy, and significant areas of accomplished or needed endeavor in the total field of residues of these and other foreign chemicals in any segment of the environment. These reviews are either general or specific, but properly they may lie in the domains of analytical chemistry and its methodology, biochemistry, human and animal medicine, legislation, pharmacology, physiology, regulation, and toxicology; certain affairs in the realm of food technology concerned specifically with pesticide and other food-additive problems are also appropriate subject matter. The justification for the preparation of any review for this book-series is that it deals with some aspect of the many real problems arising from the presence of any "foreign" chemicals in our surroundings. Thus, manuscripts may encompass those matters, in any country, which are involved in allowing pesticide and other plant-protecting chemicals to be used safely in producing, storing, and shipping crops. Added plant or animal pest-control chemicals or their metabolites that may persist into meat and other edible animal products (milk and milk products, eggs, etc.) are also residues and are within this scope. The so-called food additives (substances deliberately added to foods for flavor, odor, appearance, etc., as well as those inadvertently added during manufacture, packaging, distribution, storage, etc.) are also considered suitable review material. In addition, contaminant chemicals added in any manner to air, water, soil or plant or animal life are within this purview and these objectives.

Manuscripts are normally contributed by invitation but suggested topics are welcome. Preliminary communication with the editors is necessary before volunteered reviews are submitted in manuscript form.

Department of Entomology
University of California
Riverside, California
March 30, 1983

F.A.G.
J.D.G.

Table of Contents

Fate of polynuclear aromatic compounds (PNAs) in soil-plant systems

By

R. C. SIMS* AND M. R. OVERCASH**

Contents

---

*Department of Civil and Environmental Engineering, Utah State University, and Utah Water Research Laboratory, Logan, UT 84322.

**Department of Chemical Engineering, North Carolina State University, Raleigh, NC 27650.

Residue Reviews, Volume 88.

I. Introduction

a) Chemical structure, properties, production, sources, and historical perspectives

Polycyclic aromatic hydrocarbons (PAH) consist of three or more fused benzene rings in linear, angular, or cluster arrangements. Substitution of carbon in the benzene ring with nitrogen, sulfur, oxygen, or other elements creates heterocyclic aromatic compounds (Blumer 1976). Excluding diphenyl types there are some 70 possible isomers of 4-6 fused rings. Additionally these isomers may be substituted by a variety of substituents (Erskine and Whitehead 1975). The aromatic hydrocarbons and heterocycles, unsubstituted and substituted, are referred to alternatively as polynuclear aromatics (PNAs), polycyclic aromatic compounds (PCAs), and polycyclic organic matter (POM) (NAS 1972, U.S. EPA 1975 b).

PNAs are geochemically stable compounds which tend to be preserved in sediments and sedimentary rocks. Relative stability among PNAs is indicated by ring arrangement. Ring patterns and relative stabilities are indicated in Table I (Blumer 1976).

Structures and physical-chemical properties for several PNAs are given in Table II. The PNAs described are those listed in the U.S. Environmental Protection Agency's Consent Decree Priority Pollutant list (U.S. EPA 1976).

Production of PNAs is due to both natural and man-made processes. Table III summarizes the various sources for the production of PNAs in the environment. Table IV presents quantitative information concerning the emissions to the environment of the most studied PNA, benz(a)pyrene (B(a)P). It is apparent from Tables III and IV that anthropogenic sources are the most significant sources for PNA production and introduction into the environment.

It has been further determined (Blumer and Youngblood 1975, Giger and Blumer 1974) that the temperatures associated with pyrolysis of organic compounds determine the types of aromatic compounds that are produced. Table V summarizes the effect of temperature on extent of alkyl substitution.

With petroleum, although the PNA level is relatively low (0.4 to 1.4 ppm B(a)P), catalytic cracking processes are known to increase the PNAs which will

Table I. *Ring arrangement and relative stability* (Blumer 1976).

Ring Arrangement	linear	cluster	angular
Description	all rings in line	at least one ring surrounded on three sides	rings in steps
Stability	least	intermediate	most
Examples	anthracene tetracene	pyrene benzopyrene	phenanthrene chrysene

Table II. *Structures and physical-chemical properties of PNAs.*

PNA	Structure (no. of rings)	mp[a] °C	bp[b] °C	Aqueous solubility (mg/L)	log K_p[c]	Vapor pressure (torr at 20°C)
Naphthalene	2	80	218	30	3.37	4.92×10^{-2}
Acenaphthene	3	96	279	3.47	4.33	2.0×10^{-2}
Acenaphthylene	3	92	265	3.93	4.07	2.9×10^{-2}
Anthracene	3	216	340	0.07	4.45	1.96×10^{-4}
Phenanthrene	3	101	340	1.29	4.46	6.80×10^{-4}
Fluorene	3	116	293	1.98	4.18	1.3×10^{-2}
Fluoranthene	4	111	–	0.26	5.33	6.0×10^{-6}
Benz(a)anthracene	4	158	400	0.014	5.61	5.0×10^{-9}
Chrysene	4	255	–	0.002	5.61	6.3×10^{-7}
Pyrene	4	149	360	0.14	5.32	6.85×10^{-7}
Benz(a)pyrene	5	179	496	0.0038	6.04	5.0×10^{-7}
Benzo(b)fluoranthene	5	167	–	0.0012	6.57	5.0×10^{-7}
Benzo(k)fluoranthene	5	217	480	0.00055	6.84	5.0×10^{-7}
Dibenz(a,h)-anthracene	5	262	–	0.0005	5.97	1.0×10^{-10}
Benzo(ghi)perylene	6	222	–	0.00026	7.23	1.0×10^{-10}
Indeno(1,2,3-cd)-pyrene	6	163	–	0.062	7.66	1.0×10^{-10}

[a] mp = melting point.
[b] bp = boiling point.
[c] log K_p = logarithm of the octanol:water partition coefficient (Vesar 1979).

concentrate in residual products (Horton *et al.* 1963). Automotive lubricant oil will typically demonstrate an increase in B(a)P from 0.03 ppm to 5.8 ppm after 1,400 miles (Graf and Winter 1968). Thus PNA levels in products and in fuel oils depend upon crude oil source and upon the processing involved.

Organic geochemical reactions also produce PNAs. The apparent transformation of α-pinene to *p*-cymene (Skrigan 1951) and of abietic acid to retene in fossil pine stumps (Skrigan 1965) indicates that dehydrogenation of aliphatic compounds to aromatic compounds is an important geochemical reaction. Laboratory dehydrogenation of lanosterol with selenium yields 1,2,8-trimethylphenanthrene indicating that diagenesis of plant sterols would have given rise to this hydrocarbon (Mair and Martinez-Pico 1962).

Anthropogenic combustion of fossil fuels including wood, coal, and oil is believed to be the major source of PNAs identified in sediments in Massachusetts (Hites *et al.* 1977). Increased use of U.S. coal deposits by converting this resource to synthetic oil and gas products, termed synfuels, will greatly increase the production of PNAs and their possible introduction into the multimedia environment (Jackson 1979, Braunstein *et al.* 1977).

Table III. *Production of PNAs in the environment.*

Anthropogenic		Non-anthropogenic	
I. Industry	Reference	I. Geochemistry	Reference
fossil fuels (gasification, liquefaction)	Hites *et al.* (1977)	coal	Diehl *et al.* (1967)
	Jackson (1979)		Hayatsu *et al.* (1975)
	Braunstein *et al.* (1977)		Braunstein *et al.* (1977)
heating and power	NAS (1972)	curtisite	Geissman *et al.* (1967)
	Suess (1976)	sedimentary rocks	Hodgson *et al.* (1968)
coke production	Suess (1976)	minerals associated with mercury ores	Blumer (1975)
	Griest & Herbes (1978)		
catalytic cracking	Ershova (1971)	volcanic activity	Suess (1976)
	Suess (1976)		Il'Nitskii *et al.* (1977)
	Tye *et al.* (1966)	minerals	Murdoch & Geissman (1967)
carbon black	NAS (1972)		
asphalt hot road mix processes	NAS (1972)	geochemical reactions of biogenic molecules	Eglinton & Murphy (1968)
	Sawicki *et al.* (1965)		
internal combustion engine	Colucci & Begeman (1971)		
	NAS (1972)		
coal tar pitch	NAS (1972)		
wastewater	Il'Nitskii (1966)		
food preparation	Bailey & Dungal (1958)		
	Panalaks (1976)		
II. Open burning		II. Biochemical synthesis	
refuse	Hangebrauck *et al.* (1964)	plants	Borneff *et al.* (1968 a)
tires	Hangebrauck *et al.* (1964)		Borneff *et al.* (1968 b)
grass, leaves	Junk & Ford (1980)		Grimmer & Duvel (1970)
	Youngblood & Blumer (1975)	anaerobic bacteria	Hase & Hites (1976)
forest and agriculture	Suess (1976)	bacteria	Mallet *et al.* (1967)
	Ryan & McMahon (1976)		Niaussat *et al.* (1970)
municipal incinerators	Davies *et al.* (1976)		

Table IV. *Estimated Benzo(a)pyrene emissions* (Suess 1976).

Source	B(a)P (tons/year) U.S.A.	Worldwide (excluding U.S.A.)	Worldwide
Heating and power generation using			
coal	431	1945	2376
oil	2	3	5
gas	2	1	3
wood	40	180	220
subtotal	475	2129	2604
Industrial processes			
coke production	192	841	1033
catalytic cracking	6	6	12
subtotal	198	847	1045
Refuse and open burning			
enclosed incineration			
commercial and industrial	23	46	69
other	11	22	33
open burning			
coal refuse fires	340	340	680
forest and agriculture	140	280	420
other	74	74	148
subtotal	588	762	1350
Vehicles			
trucks and buses	12	17	29
automobiles	10	6	16
subtotal	22	23	45
TOTAL	1283	3761	5044

Table V. *Effect of temperature on alkyl substitution of PNAs.*

Temperature	Source	Alkyl substitution
High (2,000°C)	internal combustion engine	few alkyl substituents
Intermediate (400°-800°C)	wood fires	abundance of alkyl side chains increases with decreasing temperature
Low (100°-150°C)	petroleum	numerous alkyl substituents

It is also recognized that a major part of PNAs in air must sediment on terrestrial systems by rain and snow (NAS 1972, Shabad 1975). Also PNAs in wastewater and aquatic systems move into sediments due to their low solubilities in aqueous systems (refer to Table II for log K_p values) (Lee *et al.* 1978, Griest and Herbes 1978). Thus soil and sediment act as ultimate sinks for PNAs in the multimedia environment.

A natural balance between the production and the degradation of PNAs existed prior to the twentieth century. PNA synthesis by endogenous factors including microorganisms and volcanic activity and by exogenous factors including man-made high temperature pyrolytic reactions and open burning were balanced by PNA destruction via photodegradation and microbial transformation. However, with increased industrial development and with increased emphasis on the utilization of fossil fuels as an alternative energy source, the natural balance is greatly disturbed with the result that the production of PNA and introduction into the environment greatly exceeds present processes for removal of PNAs (Suess 1976, Jackson 1979, Hites *et al.* 1977).

b) Distribution

PNAs are found distributed throughout the multimedia environment. Table VI summarizes the occurrence and concentrations of individual PNAs in soils and sediments worldwide. It is obvious from the table that PNA levels in soils and sediments are higher in closer proximity to anthropogenic activity (industry and residential areas). Soil samples taken in agricultural areas, far from cities, industrial plants, and highways contain B(a)P levels not more than 5 to 10 μg/kg dry weight. This level is the normal background level in U.S.S.R. soils and is suggested as the background of B(a)P in the global terrestrial environment (Shabad 1975).

Non-anthropogenic distribution of PNAs is primarily associated with fossil fuels and minerals resulting from geochemical reactions within the earth's crust. Coal is predominantly an aromatic material (Hayatsu *et al.* 1975). Table VII lists aromatic components and heterocycles found in bituminous coal (Wiser 1973). Wiser (1973) sugggests that up to 75% of the carbon present in bituminous coal is aromatic, with a multitude of five- and six-membered rings.

PNA minerals occur in association with mercury ores, bituminous substances, and hot springs gases worldwide. Curtisite, a greenish-yellow, fluorescent soft mineral, occurring at surface vents of the hot springs at Skaggs Springs, Sonoma County, California, was first described by Wright and Allen and a related mineral idrialite was reported from Ordejov, Moldavia, Czechoslovakia, by Tucek and Kourimsky (Blumer and Youngblood 1975). The mineral pendletonite is nearly pure coronene and was reported from the New Idria Mine, San Benito County, California (Murdoch and Geissman 1967). Table VIII is a summary of these PNA-mineral associations (Blumer 1975). The alkylation patterns in the homologous series of idrialite and curtisite resemble those from soils, marine sediments, and wood tars. Thus intermediate pyrolysis temperatures are indicated, which permit the survival of intermediate alkylation (Blumer 1975, Blumer and Youngblood 1975).

The ubiquitous distribution of PNAs in terrestrial systems may also be related to plant-lignin decomposition under coalification conditions. Lignin decomposition products include phenols, quinones, and polymerization products. Under the conditions of coalification, the formation of polymerization products derived from quinonoid substances may occur. This leads to a higher degree of aromatization (Flaig 1965).

In addition to the in-situ formation of PNAs in geochemical reactions and local production of PNAs from anthropogenic sources, two transportation modes are believed responsible for additional dispersion world-wide: (1) resuspension and transport of contaminated sediments found near urban areas; and (2) long-range airborne transport of urban air particulates. Data on PNA concentration gradients in Massachusetts Bay sediment and Gulf of Maine sediment (Table VI) support the concept of sediment transport; data on soils and sediments remote from human activity support the concept of airborne transport (Windsor and Hites 1979).

c) *Toxicity*

PNAs, many of which are carcinogenic, constitute one of the more significant groups of chemicals in the environment which might be of concern. Pott's observations in 1775 that scrotal cancer in chimney sweeps was the result of a component of soot was the first suggestion that cancer could be caused by environmental agents. A century later Yamagiwa and Ichikawa demonstrated the systematic induction of skin cancer in rabbits with coal tar extracts. Shale tars, used to reduce wind erosion of soil in the U.S.S.R., induced papillomas and carcinomas in mice (Shabad 1975). The search for the active substance in coal-related material resulted in the isolation and identification of benz(a)pyrene by British scientists 120 yrs after Pott's observations

At the present time there is a dearth of information concerning acute toxicity of PNAs. Acute toxicities to fish of 1-, 2-, and 3-ring aromatic hydrocarbons indicate a general 10-fold increase in toxicity, measured as 48 hr lethal concentration at which one-half the population of animals died, *i.e.*, LC_{50} (mg/L), per 40 to 50 unit increase in molecular weight. Data for benzene, toluene, and xylene indicate that methylation may decrease toxicity. No toxicity data are available for 4- and 5-ring PNAs. The idea that solubility may play a critical role in toxicity to fish is suggested by the observation that anthracene, which is 20 times less soluble than phenanthrene, is not toxic to fish even in supersaturated solutions (McKee and Wolf 1963).

While acute effects of heterocyclic analogs of aromatic hydrocarbons have not been studied in detail, thiophene is known to be 33% more toxic to sunfish than is benzene (Jones 1964) and thiophene and 2-methylthiophene are more toxic to mammals than are the benzene analogs (Herbes *et al.* 1976).

Table VI. *Distribution of PNAs in soils and sediments.*

PNA	Site: Type	Site: Location	Concentration (μg/kg, dry wt)	Comment	Reference
Benz(a)pyrene	soil	Moscow, USSR	8010	petroleum refinery	Shabad *et al.* (1969)
Benz(a)pyrene	estuarine sediment	Bay of Naples Naples, Italy	3030	Bagnoli, black mud	Bourcart & Mallet (1965)
Benz(a)pyrene	marine sediment	North Sea, off French Coast	0-1670	bottom mud	Mallet *et al.* (1963)
Benz(a)pyrene	estuarine sediment	Bay of Naples Naples, Italy	958	Torre del Greco, black mud	Bourcart & Mallet (1965)
Benz(a)pyrene	soil	Moscow, USSR	890	old area of city	Shabad *et al.* (1969)
Benz(a)pyrene	estuarine sediment	Bay of Naples Naples, Italy	562	Ischia, black mud	Bourcart & Mallet (1965)
Benz(a)pyrene	estuarine sediment	Bay of Naples Naples, Italy	533	San Biovanni, black mud	Bourcart & Mallet (1965)
Benz(a)pyrene	estuarine sediment	Buzzards Bay Mass.	370	0.5 mile from shore	Giger & Blumer (1974)
Benz(a)pyrene	soil	Moscow, USSR	270	suburban residential	Shabad *et al.* (1969)
Benz(a)pyrene	agricultural soil	France	260	Creuse (soil surface)	Binet & Mallet (1963)
Benz(a)pyrene	estuarine sediment	Buzzards Bay Mass.	75	1.3 mile from shore	Giger & Blumer (1974)
Benz(a)pyrene	agricultural soil	Ceske Budejovice Czechoslovakia	42.1	clay	Kolar *et al.* (1975)
Benz(a)pyrene	agricultural soil	Ceske Budejovice Czechoslovakia	37.7	clayey soil	Kolar *et al.* (1975)
Benz(a)pyrene	soil	Moscow, USSR	18	suburban forest	Shabad *et al.* (1969)
Benz(a)pyrene	agricultural soil	Kremze, Czechoslovakia	16.5	sandy clay	Kolar *et al.* (1975)

Benz(a)pyrene	agricultural soil	Kremze, Czechoslovakia	8.3	argillaceous sand	Kolar *et al.* (1975)
Benz(a)pyrene	forest soil	near Lake Constance, Germany	1.5-2.5	spruce, mixed, beech woods	Harrison *et al.* (1975)
Benz(a)pyrene	forest soil	Bodensee area, Germany	1.5-2.5	fir, beech, mixed conifer	Borneff & Kunte (1963)
Benz(a)pyrene	estuarine sediment	Bay of Naples Naples, Italy	1.4	Capri, sand	Bourcart & Mallet (1965)
Benz(a)pyrene	garden soil	South Bohemia	0.010-0.04	rural area	Hites *et al.* (1977)
Pyrene	estuarine sediment	Buzzards Bay Mass.	960	0.5 mile from shore	Giger & Blumer (1974)
Pyrene	estuarine sediment	Buzzards Bay Mass.	100	1.3 mile from shore	Giger & Blumer (1974)
Pyrene	forest soil	Bodensee area Germany	10-15	pine, beech, mixed conifer	Borneff & Kunte (1963)
Benz(a)anthracene	estuarine sediment	Buzzards Bay Mass.	330	0.5 mile from shore	Giger & Blumer (1974)
Benz(a)anthracene	estuarine sediment	Buzzards Bay Mass.	41	1.3 mile from shore	Giger & Blumer (1974)
Benz(a)anthracene	forest soil	Bodensee area, Germany	10	pine, beech, mixed conifer	Borneff & Kunte (1963)
Benzo(ghi)perylene	estuarine sediment	Buzzards Bay Mass.	280	0.5 mile from shore	Giger & Blumer (1974)
Benzo(ghi)perylene	estuarine sediment	Buzzards Bay Mass.	66	1.3 mile from shore	Giger & Blumer (1974)
Benzo(ghi)perylene	forest soil	near Lake Constance, Germany	10-20	spruce, beech, mixed woods	Harrison *et al.* (1975)
Perylene	estuarine sediment	Buzzards Bay Mass.	94	0.5 mile from shore	Giger & Blumer (1974)
Perylene	estuarine sediment	Buzzards Bay Mass.	26	1.3 mile from shore	Giger & Blumer (1974)

Table VI (*continued*)

PNA	Site		Concentration	Comment	Reference
	Type	Location	(μg/kg, dry wt)		
Perylene	marine sediment	Gulf of Mexico	0.1-12		Bartle *et al.* (1976)
Benz(e)pyrene	estuarine sediment	Buzzards Bay Mass.	310	0.5 mile from shore	Giger & Blumer (1974)
Benz(e)pyrene	estuarine sediment	Buzzards Bay Mass.	59	1.3 mile from shore	Giger & Blumer (1974)
Chrysene	estuarine sediment	Buzzards Bay Mass.	240	0.5 mile from shore	Giger & Blumer (1974)
Chrysene	estuarine sediment	Buzzards Bay Mass.	40	1.3 mile from shore	Giger & Blumer (1974)
Anthanthrene	estuarine sediment	Buzzards Bay Mass.	34	0.5 mile from shore	Giger & Blumer (1974)
Anthanthrene	estuarine sediment	Buzzards Bay Mass.	7	1.3 mile from shore	Giger & Blumer (1974)
Coronene	estuarine sediment	Buzzards Bay Mass.	20	0.5 mile from shore	Giger & Blumer (1974)
Coronene	estuarine sediment	Buzzards Bay Mass.	5	1.3 mile from shore	Giger & Blumer (1974)
Anthracene	estuarine sediment	Buzzards Bay Mass.	170	0.5 mile from shore	Giger & Blumer (1974)
Fluoranthene	forest soil	Bodensee area Germany	25-40	pine, beech, mixed conifer	Borneff & Kunte (1963)
Phenanthrene	estuarine sediment	Buzzards Bay Mass.	33	1.3 mile from shore	Giger & Blumer (1974)

Benzo(b)fluoranthene	forest soil	Bodensee area Germany	20-30	pine, beech, mixed conifer	Borneff & Kunte (1963)
1,12-Benzperylene	forest soil	Bodensee area Germany	10-20	pine, beech, mixed conifer	Borneff & Kunte (1963)
Benzo(k)fluoranthene	forest soil	Bodensee area Germany	10-15	pine, beech, mixed conifer	Borneff & Kunte (1963)
Indeno(1,2,3-cd)-pyrene	forest soil	Bodensee area Germany	10-15	pine, beech, mixed conifer	Borneff & Kunte (1963)

Table VII. *Aromatic components and heterocycles in bituminous coal.*

I. Hydrocarbons	II. Sulfur heterocyclics	III. Nitrogen heterocyclics	IV. Oxygen heterocyclics
benzene	benzothiophene	pyridine	benzofuran
naphthalene	dibenzothiophene	quinoline	dibenzofuran
phenanthrene		carbazole	
fluoranthene			

Subacute effects of PNAs, including carcinogenicity and mutagenicity, are of major concern and have been studied more intensively. Fish from industrially contaminated rivers have been shown to develop four times more tumors than fish in unpolluted water (Brown *et al.* 1973). Lee *et al.* (1978) demonstrated a rapid bioaccumulation of PNAs including naphthalene, anthracene, fluoranthene, benz(a)anthracene, and benz(a)pyrene by oysters. The rate of excretion of these PNAs after transfer to uncontaminated water was inversely proportional to molecular size. Calculated half-lives assuming exponential discharge from the oysters ranged from 2 days for naphthalene to 18 days for B(a)P. The ability of PNAs to bioaccumulate in living organisms was also demonstrated by Southworth *et al.* (1978). PNAs were concentrated from a high of 10,000-fold for benz(a)anthracene to a low of 100-fold for naphthalene with *Daphnia pulex* test organisms.

PNAs can be taken into the body by inhalation, skin contact, or ingestion, although they are poorly absorbed from the gastrointestinal tract. The main routes of elimination of both PNAs and their metabolites include the hepatobiliary system and the GI tract (Braunstein *et al.* 1977).

Research into the biochemical mechanisms of action of chemical carcinogens is the result of development of sensitive bioassays and analytical techniques for analyses of the highly reactive and transient intermediates formed during the metabolism of PNAs. Animal studies indicate that PNAs are metabolized by liver

Table VIII. *PNA-mineral associations in the earth's crust* (Blumer 1975).

Mineral	PNAs	Mineral	PNAs
Curtisite	benzofluorene chrysene dibenzofluorene tetrabenzothiophene coronene anthanthrene picene	Idrialite	tribenzofluorene benzopicene pentabenzothiophene tetrabenzofluorene dibenzopicene picene

mixed function oxidases to epoxides, dihydrodiols, phenols, and quinones. These reactions represent an intoxication of the parent PNA as the metabolites have been identified as the mutagenic, carcinogenic, and teratogenic agents. Electrophilic metabolites of PNAs (epoxides) have been demonstrated to bind to genetic material including DNA and RNA which represent the basis of tumor initiation (Braunstein 1977, Hodgson and Guthrie 1980, Wright 1980, Bigger *et al.* 1980, Simmon and Baden 1980).

Metabolic formation of bay region diol epoxides represents an important pathway by which PNAs are activated to carcinogens. Such metabolic activation proceeds via initial formation of the dihydrodiol with the bay region double bond, followed by subsequent oxidation of the dihydrodiol to the bay region diol epoxide. Metabolic activation of B(a)P to bay region diol epoxides is illustrated in the following scheme (Thakker *et al.* 1979) (Fig. 1).

Liver microsomal enzymes of trout fish were shown to activate B(a)P, 2-aminofluorene, and 2-acetylaminofluorene to compounds which are potent frameshift mutagens in the Ames *Salmonella typhimurium*/mammalian microsome assay (Ahokas *et al.* 1976).

The effects of PNAs on terrestrial and aquatic organisms is summarized in Table IX. While Hass and Applegate (1975) demonstrate a structure-specific response of microorganisms to PNAs, Calder and Lader (1976) demonstrated that PNAs inhibited the growth of bacteria in a manner which was dose-related and a function of water solubility. Saturated solutions of B(a)P and naphthalene

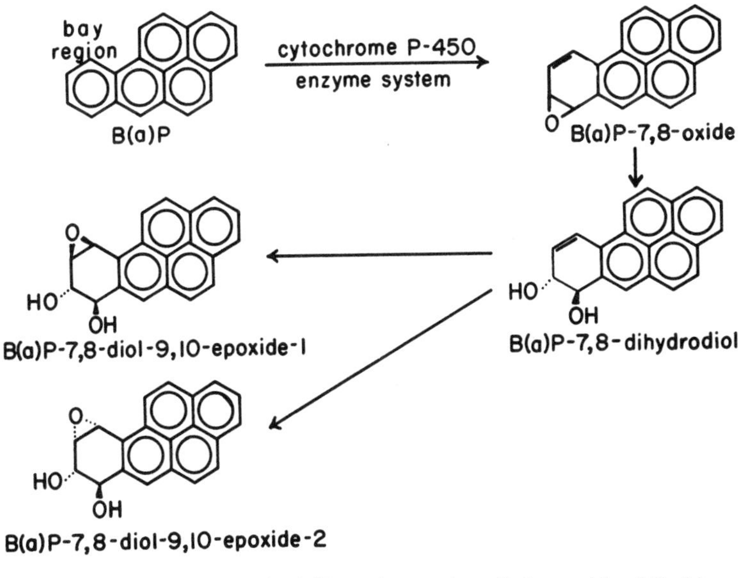

Fig. 1. Metabolic activation of B(a)P to bay region diol epoxides (Thakker *et al.* 1979).

Table IX. *Effects of PNAs on aquatic and terrestrial organisms.*

PNA	Organism	Concentration (mg/L) and/or response	Reference
Anthracene	*E. coli*	inhibit growth	Hass & Applegate (1975)
Phenanthrene	*E. coli*	inhibit growth	Hass & Applegate (1975)
Dibenz(a,c)anthracene	*E. coli*	inhibit growth	Hass & Applegate (1975)
Benz(a)anthracene	*E. coli*	stimulate growth	Hass & Applegate (1975)
Dibenz(a,h)anthracene	*E. coli*	stimulate growth	Hass & Applegate (1975)
Fluoranthene	*C. vulgaris, S. obliquus, A. oranunii*	stimulate growth at 0.01-0.02 ppm	Graf & Nowak (1966)
1,12-benzoperylene	*C. vulgaris, S. obliquus, A. oranunii*	stimulate growth at 0.01-0.02 ppm	Graf & Nowak (1966)
Indeno (1,2,3-cd) pyrene	*C. vulgaris, S. obliquus, A. oranunii*	stimulate growth at 0.01-0.02 ppm	Graf & Nowak (1966)
Benz(a)pyrene	*C. vulgaris, S. obliquus, A. oranunii*	stimulate growth at 0.01-0.02 ppm	Graf & Nowak (1966)
Dibenz(a,h)anthracene	sludge microorganisms	inhibit respiration	Malaney *et al.* (1967)
Dibenz(a,h)acridine	sludge microorganisms	inhibit respiration	Malaney *et al.* (1967)
Dibenz(a,j)acridine	sludge microorganisms	inhibit respiration	Malaney *et al.* (1967)
7-methylbenz(a) anthracene	sludge microorganisms	inhibit respiration	Malaney *et al.* (1967)
Benz(a)pyrene	newt	tumor induction	Arffman & Christensen (1961)
Benz(a)pyrene	clawed toad (*Xenopus laevis*)	tumor induction	Balls (1964)
Pyrene	chinook salmon	1.0 (lethal in 9-13 hr)	Macphee & Ruelle (1969)
3-methylcholanthrene	chinook and Coho salmon	10 (no effect)	Macphee & Ruelle (1969)
Pyridine	fish	15 (nervous system)	McKee & Wolf (1963)
Quinoline	trout	5.0 (lethal in 14 hr)	McKee & Woff (1963)
Quinoline	ciliates	750 (lethal)	McKee & Wolf (1963)

had the same impact on growth. Thus when evaluating the effects of PNAs in terrestrial and aquatic systems, it is important to balance inherent toxicity on a molar basis against water solubility (U.S. EPA 1978).

Acute and chronic toxicities for PNAs included in the U.S. Environmental Protection Agency's Consent Decree Priority Pollutant List are summarized in Table X. Values shown are obtained from standard toxicity testing with animals for acute LD_{50} (concentration in animal tissue, mg/kg, at which one half of the population died) toxicity and for carcinogenicity; values for mutagenicity are based on the standardized Ames *Salmonella typhimurium*/mammalian microsomal assay.

An essential concept resulting from occupational exposure, animal experiments, and ecological observations is the dose-response relationship in carcinogenesis. In general, a decrease of the dose leads to a decrease of the number of induced tumors and to the delay of their appearance. With the ubiquitous occurrence and distribution of PNAs in the multimedia environment, therefore, it is important to investigate the behavior and fate of PNAs in the environment and to establish non-toxic levels for these compounds in the terrestrial environment.

d) Objectives

Many PNAs are known to be carcinogenic. Therefore a better understanding of their formation, behavior, and fate in the environment is necessary. PNAs are

Table X. *Acute and chronic toxicities of PNAs* (Kingsbury *et al.* 1979, McCann and Ames 1975).

	Acute toxicity	Chronic toxicity	
PNA	LD_{50} (oral, rat or oral, mouse) (mg/kg)	Carcinogenicity in animals (mg/kg)	Mutagenicity (Ames Assay)
Naphthalene	1,780	negative	negative
Acenaphthene	unknown	unknown	unknown
Acenaphthylene	unknown	unknown	unknown
Anthracene	unknown	3,300	negative
Phenanthrene	700	negative	negative
Fluorene	unknown	unknown	unknown
Fluoranthene	2,000	unknown	positive
Benz(a)anthracene	unknown	2	positive
Chrysene	unknown	99	positive
Pyrene	unknown	negative	negative
Benz(a)pyrene	50	2 μg/kg	positive
Benzo(b)fluoranthene	unknown	40	unknown
Benzo(k)fluoranthene	unknown	72	unknown
Dibenz(a,h)anthracene	unknown	6 μg/kg	positive
Benzo(ghi)perylene	unknown	unknown	unknown
Indeno(1,2,3-cd)pyrene	unknown	72	unknown

relatively difficult to remove from wastewaters, are slowly degraded, possess potential for bioaccumulation, carcinogenicity and mutagenicity, and their behavior in the environment at trace concentrations is poorly understood (Herbes *et al.* 1976).

This review was prepared to evaluate PNA decomposition and stability in the soil-plant environment. Factors involved in microbial transformation and decomposition and in plant uptake and accumulation of PNAs as influenced by environmental conditions were addressed. State-of-the-art knowledge on the interactions of PNAs with the soil-plant system are presented. Specific objectives of this review include the identification of factors and problems associated with agronomic practices for managing wastes containing PNAs by land application at assimilative rates that are safe and economically acceptable. Factors affecting the behavior and fate of PNAs in soil and in crops of agronomic importance have been elucidated.

II. Microbial decomposition and stability of PNAs in soils

In spite of the stability of the aromatic nucleus, due to the large negative resonance energy, microorganisms exist which can degrade the benzene ring (Gibson 1968). Aerobic bacteria possess the unique biochemical ability to catalyze the oxidation of compounds using molecular oxygen to initiate reaction sequences that enter central pathways of metabolism including the Krebs cycle and the fatty acid spiral (Dagley 1975). In the mixed microbial population of soil one type of microorganism may partially metabolize a compound without deriving carbon or energy for cell synthesis, but it may furnish a suitable growth substrate for another microorganism. Shabad (1975) suggested that the main role in the degradation of PNAs is played by soil bacteria.

While there has been a considerable amount of research conducted in Europe and the U.S.S.R. concerning PNA-soil interactions, in the U.S. there has been an emphasis on investigating metabolic pathways for PNA degradation and transformation. Until 1971, research emphasis was placed on the pathways and mechanisms utilized by microorganisms to metabolize oxygenated aromatic compounds, while parent aromatic compounds received much less attention (Gibson 1971). As a result of intensified efforts directed toward microbial metabolism of PNAs, much is known at present concerning the microbial metabolism (including bacteria and fungi) of benzene, naphthalene, anthracene, phenanthrene, and benz(a)pyrene. However, there is a paucity of information on microbial metabolism of other PNAs.

a) Mechanisms and products of biodecomposition

Four intensive areas of research concerning the mechanisms and products of microbial decomposition of PNAs include: (1) initial reactions including hydroxylations and ring cleavage, (2) metabolic intermediates in degradation, (3) enzyme regulation in metabolism, and (4) cooxidation reactions. Each of these areas will be discussed individually.

The unsubstituted saturated ring structure of PNAs is biologically inert. Dihydroxylation of the aromatic nucleus is a prerequisite for enzymatic cleavage of the benzene ring by oxygenases that incorporate molecular oxygen into the ring (Dagley 1975). The hydroxyl groups may be *ortho* to each other, as in catechol and protocatechuic acid, or *para* to each other as in gentisic and homogentisic acids (Gibson 1968). Thus the incorporation of molecular oxygen into a PNA is an activation step that makes the molecule biologically reactive and also a substrate for ring fission.

Substituted aromatic ring structures may be initially metabolized via the substituted group or the aromatic ring. While phenyl-substituted acids are degraded by *Nocardia* sp. through beta-oxidation of the side chain (Webley *et al.* 1962), *Pseudomonas putida* initiates oxidation of toluene and isopropylbenzene with the aromatic nucleus (Gibson 1968).

The initial reactions in the activation and oxidation of PNAs involve oxygenases. These are enzymes which catalyze oxygen-fixing reactions. Two subclasses of oxygenases include monooxygenases and dioxygenases. Monooxygenases, also called mixed function oxidases, are characteristic of eucaryotes including mammals and fungi. Monooxygenases catalyze the incorporation of a single atom of oxygen into a substrate. Dioxygenases are characteristic of procaryotes including bacteria, and catalyze the incorporation of both atoms of molecular oxygen into a substrate. Figure 2 illustrates the initial reactions of oxygenases on PNAs and the initial products for eucaryotes and procaryotes.

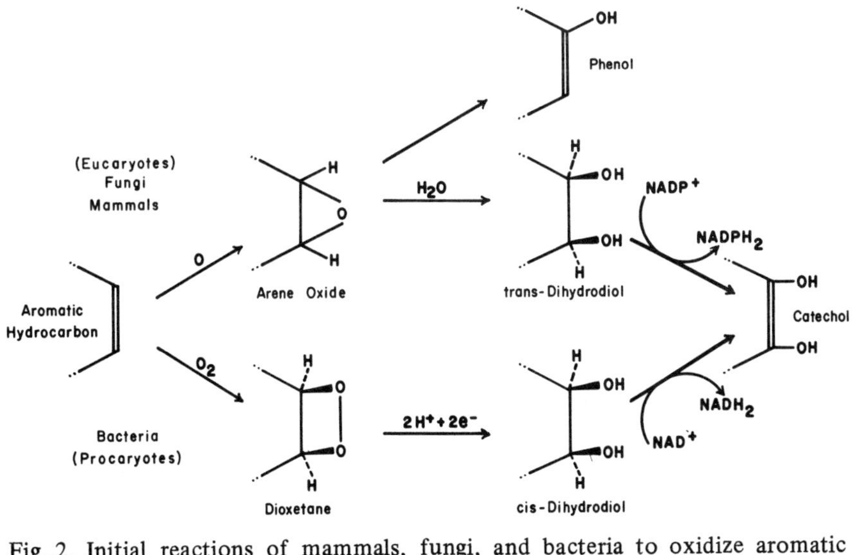

Fig. 2. Initial reactions of mammals, fungi, and bacteria to oxidize aromatic hydrocarbons (Cerniglia *et al.* 1979, Gibson 1976).

Table XI. *Eucaryotic vs. procaryotic metabolism of PNAs.*

Organism	Metabolizing system	Substrate specificity	Oxygen (O_2) incorporation	Products
Eucaryotes (mammals) (fungi)	cytochrome P-450 mixed function oxidase (monooxygenase)	general	one atom	Arene oxides *trans*-dihydrodiols
Procaryotes (bacteria)	dioxygenases	specific	both atoms	Dioxetanes *cis*-dihydrodiols

It was pointed out by Gibson (1971) that the stereochemistry of the dihydrodiols is of key importance in indicating the fundamental mechanisms by which organisms metabolize aromatic compounds. These fundamental mechanisms of metabolism are directly related to the mutagenic properties of metabolized PNAs. Bacterial oxidation or aromatic hydrocarbons is demonstrated to proceed via a dioxygenase-catalyzed reaction in which the substrate is converted to a *cis*-dihydrodiol containing both atoms of O_2. It was also demonstrated that dioxetanes are intermediates for reductive generation of *cis*-stereochemistry in the diol. Consequently, mammalian and fungal metabolism of aromatic hydrocarbons occurs via a monooxygenase-catalyzed reaction whereby the *trans*-dihydrodiol containing only one of the atoms of O_2 is produced. The *trans*-stereochemistry in the diol indicates enzymatic hydration of an intermediate arene oxide (Jerina *et al.* 1971). It is the arene oxide (epoxide) forms of PNAs which, due to their electrophilic nature, bind to nucleophilic sites including DNA and RNA for tumor initiation in carcinogenesis (Thakker *et al.* 1979, Gibson 1976, Braunstein 1977).

Table XI summarizes the eucaryotic *versus* procaryotic metabolism of PNAs (Dagley 1975, Gibson 1976).

Differences in the mechanism of aromatic compound metabolism by mammals and bacteria have important implications for engineering techniques for controlling and possibly detoxifying PNAs in waste streams. Ames *et al.* (1972) have demonstrated that the epoxides formed by mammalian microsomal mixed function oxidation (metabolism) of PNAs function as frameshift mutagens. Frameshift mutagens alter the reading frame during transcription of genetic information from deoxyribonucleic acid to ribonucleic acid resulting in the synthesis of faulty proteins. It has been suggested by this investigator and by others (McKenna 1977) that since epoxides are not intermediates in the bacterial metabolism of PNAs, bacterial degradation of PNAs may provide an engineering management technique for the detoxification and assimilation of PNAs in the environment without mutagen formation.

The recent discovery that fungal metabolism of PNAs is qualitatively similar to mammalian metabolism in that arene oxides (epoxides) are produced as initial oxidation products (Cerniglia and Gibson 1979, Cerniglia *et al.* 1979) also has important implications for engineering management of wastes containing PNAs. Bacteria are less common in acid habitats. The microbial community in areas of low pH is dominated by fungi. The insensitivity of many fungi to low pH and the inhibition of bacterial activity at low pH provide the fungi with a competitive advantage for growth. As a result fungi play the major role in microbial transformations in acid habitats (Alexander 1977). Increasing pH to neutrality, however, results in the dominance of soil bacterial numbers. Thus soil pH may

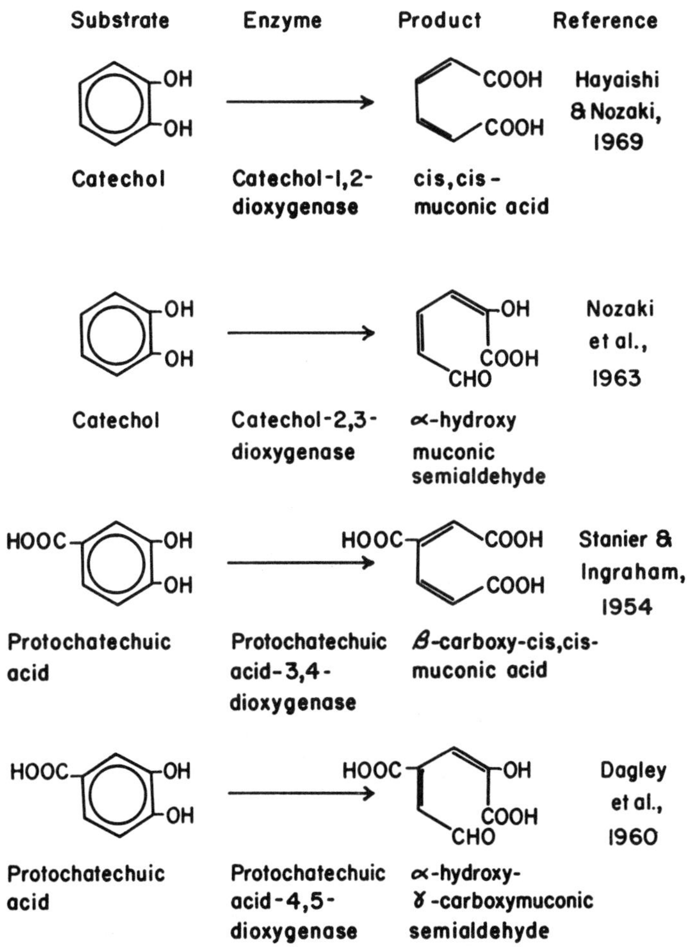

Fig. 3. Aromatic ring cleavages by oxygenase enzymes.

Table XII. *Initial oxidation products of soil microorganism metabolism of PNAs.*

Soil microorganism	Substrate or cosubstrate	Product	Reference
I. Bacteria			
Pseudomonas putida	toluene, isoproply-benzene	ortho-dihydroxy compounds	Gibson (1968)
Pseudomonas putida	naphthalene	*cis*-dihydrodiols	Jeffrey *et al.* (1975)
Pseudomonas putida	benzene	*cis*-1,2-dihydrodiols	Gibson *et al.* (1970)
Pseudomonas putida	benzene	catechol and *cis*-benzene glycol	Gibson *et al.* (1968)
Pseudomonas sp.	protochatechuic acid	α-hydroxy-β-carboxymuconic semialdehyde	Dagley *et al.* (1960)
Pseudomonas sp.	naphthalene	*cis*-dihydrodiols	Jerina *et al.* (1971)
Pseudomonas arvilla	catechol	α-hydroxymuconic semialdehyde	Nozaki *et al.* (1963)
Pseudomonas fluorescens	naphthalene	*cis*-dihydrodiols	Jeffrey *et al.* (1975)
Pseudomonas aeruginosa	phenanthrene	1-hydroxy-2-naphtholic acid	Rogoff & Wender (1957)
Beijerinckia sp.	benz(a)pyrene	*cis*-9,10-dihydrodiol	Gibson *et al.* (1975)
Beijerinckia sp.	benz(a)anthracene	*cis*-1,2-dihydrodiol	Gibson *et al.* (1975)
Beijerinckia sp.	phenanthrene	*cis*-dihydrodiols	Selander *et al.* (1971)
Beijerinckia sp.	anthracene	*cis*-1,2-dihydrodiol	Akhtar *et al.* (1975)
Rhizobium sp.	catechol	cleavage products	Hussein *et al.* (1974)
Rhizobium sp.	protocatechuic acid	cleavage products	Hussein *et al.* (1974)
Rhizobium sp.	*p*-hydroxybenzoic acid	protocatechuic acid	Hussein *et al.* (1974)
Rhizobium sp.	salicyclic acid	gentisic acid	Hussein *et al.* (1974)
II. Actinomycetes			
Nocardia sp.	phenyl-substituted acids	β-oxidation of side chain	Webley *et al.* (1962)
Nocardia sp.	*p*-xylene	α,α-dimethylmuconic acid	Jamison *et al.* (1969)

III. Fungi			
Cunninghamella elegans	benz(a)pyrene	*trans*-9,10-dihydrodiol	Cerniglia & Gibson (1979)
		trans-7,8-dihydrodiol	
		B(a)P-1,6-quinone	
		B(a)P-3,6-quinone	
		9-hydroxy-B(a)P	
		3-hydroxy-B(a)P	
Cunninghamella elegans	naphthalene	*trans*-1,2-dihydrodiol	Cerniglia & Gibson (1978)
		1-naphthol	
Cunninghamella elegans	dibenzofuran	*trans*-2,3-dihydrodiol	Cerniglia *et al.* (1979)
		2-hydroxydibenzofuran	
		3-hydroxydibenzofuran	
		2,3-dihydroxydibenzofuran	
Cunninghamella elegans	naphthalene	*trans*-naphthalene diol	Cerniglia *et al.* (1979)
		4-hydroxy-1-tetralone	
		1,2-naphthoquinone	
		1,4-naphthoquinone	
		1-naphthol	
		2-naphthol	
Saccharomyces cerevisiae	naphthalene	*trans*-1,2-dihydrodiol	Cerniglia *et al.* (1979)
		1-naphthol	
		2-naphthol	

be an important engineering variable for the control of PNA metabolism in terrestrial environments.

Major reactions catalyzed by dioxygenases include cleavage of the aromatic double bond which may be located (1) between two hydroxylated carbon atoms, (2) adjacent to a hydroxylated carbon atom, or (3) in an indole ring (Hayaishi and Nozaki 1969).

Catechol and protocatechuic acid are precursors to ring cleavage in the microbiol degradation of many PNAs. Figure 3 illustrates the enzymes and products formed in microbial metabolism of these substrates.

Initial oxidation products catalyzed by microbial oxygenases including ring hydroxylation and fission are summarized in Table XII. Cerniglia and Gibson (1979) demonstrated for the first time that a eucaryotic microorganism can oxidize the potent carcinogen benz(a)pyrene to metabolites qualitatively similar to those produced by mammalian microsomes. Cerniglia *et al.* (1979) also demonstrated that the fungal metabolism of dibenzofuran, an oxygen containing heterocycle, proceeds through the DNA-binding-arene-oxide pathway. It is evident from Table XII and from the previous discussion that the soil microbial community possesses two different metabolic pathways for the initial oxidation of PNAs.

Following initial hydroxylation and ring fission of the aromatic nucleus, the soil microbial community transforms the phenols and carboxylic acids produced through well-known pathways. Figure 4 illustrates the microbial degradation of naphthalene by *Pseudomonas* sp. It is evident that microorganisms metabolize

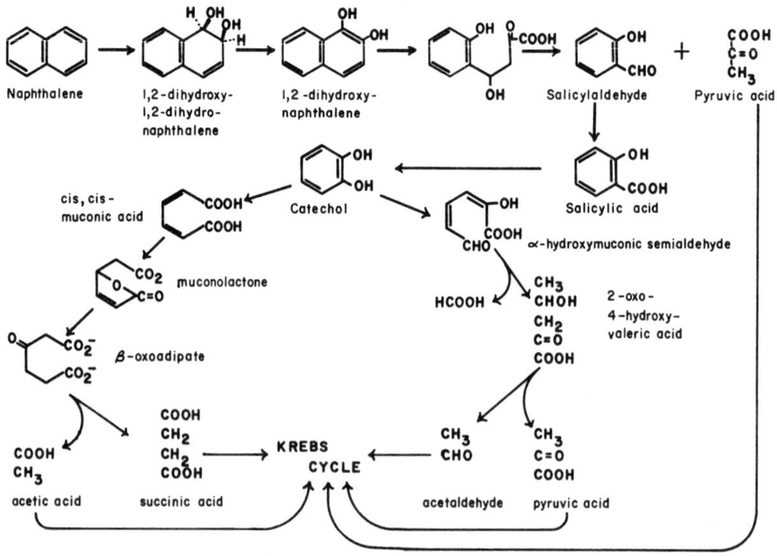

Fig. 4. Metabolism of naphthalene by soil bacteria (Dean-Raymond and Bartha 1975, Gibson, 1976, Gibson 1968).

PNAs one cyclic unit at a time. *Pseudomonas* and *Mycobacterium*, among other microorganisms, oxidize anthracene and phenanthrene to intermediates with only two rings with further oxidation to catechol and to Krebs cycle intermediates (Alexander 1977). Pathways for creating Krebs cycle intermediates from PNA degradation are presented in Figure 4.

Little is known, however, concerning the intermediates and breakdown products for PNAs containing greater than three rings. Attempts to isolate microorganisms that utilize these compounds as sole sources of carbon and energy have not been successful (Gibson 1976). Cooxidation, the fourth topic to be considered, may play an important role in the degradation of these PNAs.

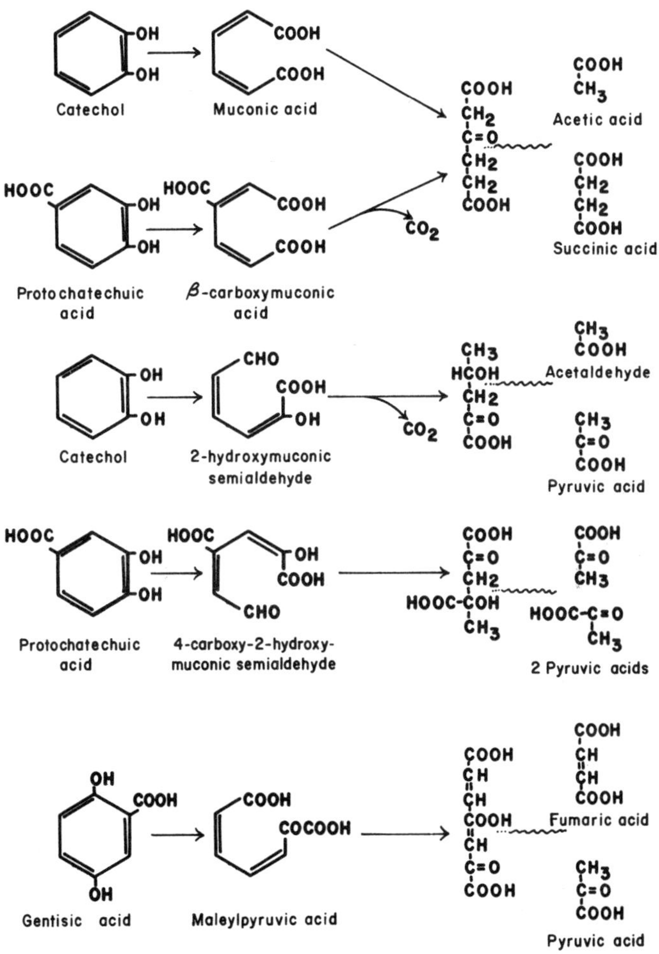

Fig. 5. Intermediates in the degradation of PNAs by soil microorganisms (Alexander 1977).

Although initial reactions in the degradation of PNAs by soil microorganisms differ considerably, only a few initial intermediates are produced including catechol, protocatechuic acid and gentisic acid. These three compounds are degraded by five similar pathways to produce succinic, fumaric, pyruvic, and acetic acids, and acetaldehyde, all of which are readily utilized for cell synthesis and energy by microbial cells (Alexander 1977). These five cellular intermediates are presented in Figure 5.

The mechanisms of enzyme regulation in the degradation of PNAs is the third important area of research. All the enzymes for the reactions discussed have been shown to be inducible. Cells exposed to one metabolite synthesize large amounts of enzymes to degrade that metabolite (Dagley 1975).

The regulation of naphthalene metabolism in soil pseudomonads has been well characterized. Until 1974, although the metabolic pathway for naphthalene degradation was characterized, the mechanism of enzymatic regulation of the initial steps had not been elucidated. Shamsuzzaman and Barnsley (1974) demonstrated that naphthalene oxygenase was induced in soil pseudomonads when the organisms were grown in salicylate as a carbon and energy source, or when salicylate was added to cultures growing on succinate. Two structural analogues of salicylate, 2-aminobenzoic acid and 2-hydroxylbenzyl alcohol, also induced naphthalene oxygenase. From these observations it became apparent that salicylate was probably the inducer of naphthalene oxygenase. The same researchers also reported the induction of naphthalene oxygenase, 1,2-dihydroxynaphthalene oxygenase, and salicylaldehyde dehydrogenase with salicylate. They suggested that all of the enzymes converting naphthalene to salicylate are induced and regulated coordinately.

There does appear to be considerable agreement in the literature with respect to the need for induction of microbial enzymes in order to achieve significant metabolism of PNAs. Shabad (1967) demonstrated that need for previously acclimated B(a)P-metabolizing microorganisms to be used to degrade newly added B(a)P to soil systems. Cultivation of B(a)P-acclimated cells in a medium free of B(a)P resulted in decreased capacity to degrade B(a)P, while addition of B(a)P to the medium restored the degradation capacity of cells. Poglazova *et al.* (1967 a, 1968) found that bacteria isolated from soil containing large amounts of B(a)P were much more active in degrading B(a)P than bacteria isolated from uncontaminated soils.

Cooxidation of aromatic compounds is the fourth area of interest in PNA degradation. It has been shown by Tausson (1950) that naphthalene, anthracene, and phenanthrene are degraded by microorganisms to salicylic acid, then into the catechol pathways described previously. Thus these PNAs can serve as sources of carbon and energy for cell synthesis. Little is known, however, about the microbial degradation of PNAs with more than three rings. No organisms have been isolated which can grow on PNAs possessing more than three fused rings (McKenna 1977).

Table XIII. *Rate of oxidation of PNAs by soil bacteria (bioconversion)* (McKenna 1977).

PNA	Rate of oxidation: *Pseudomonas putida* Naphthalene = 100%	Rate of oxidation: *Flavobacterium* sp. Phenanthrene = 100%
Naphthalene	100	79.8
Anthracene	9.8	36.0
Phenanthrene	67.0	100
1,2-Benzanthracene	0	10.1
2,3-Benzanthracene	2.7	1.1
Chrysene	0	7.9
Pyrene	1.1	10.1
Triphenylene	0	0
Benz(a)pyrene	0	0
Dibenz(a,h)anthracene	0	0
Perylene	0	0

McKenna (1977) investigated the enzymatic conversion in PNAs by nonproliferating cellular suspensions of soil bacteria grown on naphthalene or phenanthrene. This type of reaction is properly identified as bioconversion, rather than cooxidation, because there is no growth substrate present when the cells are oxidizing PNAs (Perry 1979). Enzymes of broad specificity and capable of PNA conversion were not found for PNAs containing more than three rings as shown in Table XIII.

Cooxidation studies, *i.e.*, measurement of oxidation of nongrowth substrates during growth of a microorganism on a carbon and energy source, were also employed using pure cultures of soil bacteria (McKenna 1977). *P. putida* was unable to oxidize pyrene as a cosubstrate, whereas *Flavobacterium* sp. oxidized 30% in 2 wk in the presence of phenanthrene as growth substrate. *P. putida* was unable to metabolize benz(a)pyrene as a cosubstrate, whereas *Flavobacterium* sp. degraded more than 50% in the 4. *Flavobacterium* also cometabolized 37% of benz(a)anthracene added and 62% of the dibenz(a,h)anthracene. It appears from these results that cooxidation may be an important mechanism whereby soil microorganisms degrade PNAs in the terrestrial environment.

In another cooxidation experiment (McKenna 1977) with a mixed microbial culture obtained from a polluted stream in Urbana, Illinois, the ability of microorganisms to degrade multi-ring PNAs was again demonstrated. Results are presented in Table XIV. It is obvious that four- and five-ring PNAs which cannot support microbial growth can nevertheless be degraded by microorganisms when the PNAs are present as cosubstrates.

Cooxidation of one- and two-ring aromatic compounds by soil microorganisms is very common. *Nocardia* sp. growing on n-hexadecane readily cooxidized

Table XIV. *Cooxidation of PNAs by a mixed microbial population* (McKenna 1977).

Non-growth substrate	Growth substrate	Amount non-growth substrate remaining after 4 wk (%)
Pyrene	naphthalene	36.7
	phenanthrene	47.2
Benz(a)pyrene	naphthalene	83.5
	phenanthrene	36.3
Benz(a)anthracene	naphthalene	58.3
	phenanthrene	33.8
Dibenz(a,h)anthracene	naphthalene	92.7
	phenanthrene	32.9

methyl-substituted mono- and dicyclic aromatic hydrocarbons to hydroxylated and acidic products (Raymond *et al.* 1967). Ring cleavage of *p*-xylene by *Nocardia coralina* V-49 growing on *n*-hexadecane as growth substrate was demonstrated by Jamison *et al.* (1969).

b) Rates of decomposition

Degradation rates and extent of degradation of PNAs and of other organic chemicals in soil systems are influenced by environmental factors, microbial flora, and physicochemical properties of the organic chemical. Important environmental factors include temperature, pH, oxygen status, presence of other substances that may serve as substrate co-metabolites, and soil type, history, moisture, and nutrient status. Microbial factors include acclimation status and populations present as well as relative proportions of bacteria, fungi, and actinomycetes. Physicochemical properties of the organic chemical of interest that are important include structure, size, concentration, and lipophilicity. Several excellent reviews concerning microbial interactions in soil systems (Bazin *et al.* 1976, Hattori and Hattori 1976, Stotzky 1973), and techniques for determining persistence and routes of degradation of chemicals in soil environments (U.S. EPA 1975 a), discuss these variables in detail as they relate to general principles of microbial and chemical reactions in soil.

Sherrill and Sayler (1980) showed that the rate of degradation of phenanthrene in natural aquatic systems was directly related to the history of pollution of the site, phenanthrene concentration, time of incubation, and temperature. Increasing concentration 100-fold (10 to 1,000 μg/L) resulted in a 3-fold increase in the rate of phenanthrene degradation (12.1 μg/L-day vs. 43.4 μg/L-day) at 25°C. Extent of degradation was 3- to fourfold greater in previously contaminated environments. Degradation of phenanthrene increased by approximately 20% for each 5°C in temperature over the range 15° to 37°C, with a maximum degradation of 90% at 37°C (at 1,000 μg/L). The mean phenanthrene biodegradation was approximately 80% after 4 wk of incubation of a 1,000 μg/L solution.

Gardner *et al.* (1979) studied the biodegradation of anthracene, fluoranthene, benz(a)anthracene, and benz(a)pyrene in fine sand, medium sand, and marsh sediment. Results are presented in Table XV. The addition of polychaete worm, *Capitella capitata*, generally increased biodegradation rates by a factor of 1.1 to 2.4.

The results shown in Table XV show that the rates of biodegradation generally were similar and higher for anthracene and fluoranthene; similar and lower for benz(a)anthracene and benz(a)pyrene. Also the % removed/wk appeared to remain relatively constant while the initial concentrations increased 2- to threefold. The rate of degradation appeared to be a function of the initial concentration.

Herbes (1981) investigated the microbial transformations of naphthalene, anthracene, benz(a)anthracene, benz(a)pyrene, and dibenz(a,h)anthracene. Sediment samples at 21°C were incubated with each radiolabeled PNA at approximately 7 μg/g for one to 24 hr. Mean rate constants in sediment collected downstream from a coal-coking plant wastewater treatment discharge point were 7.8×10^{-2}/hr (naphthalene), 1.6×10^{-2}/hr (anthracene), and 3.3×10^{-3}/hr (benz(a)anthracene). Transformations of benz(a)pyrene and dibenz-(a,h)anthracene were not observed. In parallel water samples, only naphthalene and anthracene were observed to be degraded, with rate constants consistently 5- to 20-fold lower, respectively, than in sediment samples.

Metabolic products measured included $^{14}CO_2$, cell-bound ^{14}C, and polar products. Whereas $^{14}CO_2$ and bound ^{14}C were major transformation products in sediment assays, soluble polar ^{14}C was predominant in water samples. Non-extractable bound ^{14}C was the major transformation product of anthracene in sediments. The accumulation of ^{14}C-labeled polar metabolites separable by thin-layer chromatography was not quantitatively significant in anthracene and benz(a)anthracene assays.

Table XV. *Biodegradation of PNAs in sand and marsh sediments* (Gardner *et al.* 1979).

	PNA							
	Anthracene		Fluoranthene		Benz(a)-anthracene		Benz(a)pyrene	
Soil	C_o[a]	k[b]	C_o	k	C_o	k	C_o	k
Fine sand	13.7	2.0	18.8	1.9	20.8	1.5	9.5	1.2
Medium sand	10.3	2.4	16.5	2.4	17.2	1.8	7.6	1.4
Marsh sediment	40.0	3.0	44.5	2.0	42.8	1.4	17.9	0.84

[a] C_o = initial concentration (μg/g dry sediment).
[b] k = % PNA removed/week.

Poglazova *et al.* (1967 b) demonstrated very poor transformations of B(a)P by liquid cultures of *Bacillus megaterium* after 4 days incubation. One strain (N 13) transformed 53% B(a)P in 4 days. They showed that soil bacteria cultivated on agar which did not contain aromatic hydrocarbons lose the ability to destroy B(a)P. With soil inoculation of B(a)P and *Bacillus megaterium,* a strain not cultivated on B(a)P transformed 66% in 8 days, while a strain cultivated in the presence of B(a)P transformed 82% during the same time period.

Differences in the rates of transformation between soils and aqueous systems were noted. All of the investigated strains destroyed B(a)P to a greater extent in soil than in meat infusion agar. Strain N 5, which manifested no activity toward B(a)P in liquid culture, was very active in degrading the PNA in the contaminated soil from which it was isolated (Poglazova *et al.* 1967 b).

Poglazova *et al.* (1967 b) also demonstrated the oxidation of B(a)P in liquid cultures by *Mycobacterium* and a non-spore forming mutant of *Bacillus megaterium*. The great differences between strains of *Bacillus megaterium, Mycobacteria*, and mutant *Bacillus megaterium* suggest that the ability of microorganisms to transform B(a)P is a function of the strain and not the species.

Bacterial inoculation of sterile soil containing 20 μg B(a)P/g soil achieved significant reductions in the PNA concentration. Reduction of B(a)P ranged from 55 to 60% over the 8 day incubation period for pure cultures and increased to 84% when the two single strains PBK 13 and PBK 5 were combined in soil (Poglazova *et al.* 1967 a).

Poglazova *et al.* (1968) demonstrated that the rate of B(a)P transformation in liquid media is a function of the concentration. The % destroyed after 4 days remained relatively constant (45, 46, 33%) while the initial concentration increased 23-fold (2.0, 10.0, 46.0 μg/ml, respectively).

Bacterial inoculation of *B. megaterium* and two other strains that were acclimated to high B(a)P concentrations into sterile soil containing 130 mg/kg B(a)P resulted in 80 to 85% reduction in 8 days. These figures compare with 55 to 84% reductions in inoculated soil at 20 mg/kg. Thus the percentage B(a)P transformed remained approximately stable while the concentration was increased by a factor of seven. The rate of B(a)P transformation in soils is, therefore, shown to be a function of the soil concentration of B(a)P.

Khesina *et al.* (1969) investigated the degradation of B(a)P in soils under natural conditions of temperature and soil moisture. A control soil (C) with no history of pollution, a moderately polluted soil (M), and a highly oil-contaminated soil (N) were studied. Also a control soil with B(a)P added to increase the level to that of the (N) soil was included (B). The experiment was conducted in an unheated greenhouse for 3 mon. Water was added during the first and third months only. The results are presented in Table XVI.

Results clearly show that the degradation of B(a)P can be obtained under natural conditions and with natural soil microbial populations. Under natural conditions, as with laboratory results discussed previously, the rate of transformation appears to be a function of the soil concentration for soils that are

Table XVI. *Degradation of benzo(a)pyrene by soil microorganisms* (Khesina *et al.* 1969).

Soil	C_o [a]	B(a)P content of soil: 1 mon exposure (μg/kg soil)	2 mon exposure (μg/kg soil)	3 mon exposure (μg/kg soil)	Reduction at 3 mon (%)
C[b]	9.5	7.4	7.5	6.3	33
M[c]	545	377	346	159	71
B[d]	29,200	29,300	29,000	30,100	0
N[e]	28,500	16,000	16,700	13,700	52

[a] C_o = initial soil concentration (μg/g soil).
[b] C = control soil with no history of pollution.
[c] M = moderately oil-polluted soil.
[d] B = control soil with B(a)P added to increase the level to that of the (N) soil.
[e] N = highly oil-contaminated soil.

acclimated to the presence of PNAs. Soils (M) and (N) achieved similar reductions in B(a)P over 3 mon although initial soil levels of B(a)P differed by a factor of 50. For non-acclimated soils reductions in B(a)P were very low, some soils showing no reduction in B(a)P. Thus B(a)P can be microbially degraded in soil systems under natural conditions provided that the soil has been acclimated, *i.e.*, exposed to the presence of PNAs over a time period before introduction of additional PNA. Under these conditions for degradation, the rate of transformation appears to be a function of the soil concentration of B(a)P.

The growth and viability of soil bacteria does not appear to be affected by even high levels of B(a)P (Poglazova *et al.* 1967 a). Bacteria were shown to accumulate B(a)P primarily in lipid and other intracellular inclusions.

Table XVII summarizes the degradation of PNAs in soil systems. Information from a variety of sources was normalized with respect to kinetic parameters for describing PNA degradation as a function of soil concentration. Initial concentrations, rate constants, rates of transformation, and half-lives were calculated (Sims 1982).

From information obtained in Table XVII and from laboratory experiments conducted at North Carolina State University (Sims 1982) initial rates of transformation of PNA compounds in soil as a function of initial soil concentrations assuming first order kinetics are presented in Figure 6. These data were corrected for variations in temperature using an Arrhenius equation with coefficients developed from PNA data to a temperature of 20°C, $\theta = 1.013$. Rates were normalized to μg PNA transformed/g soil-dry wt/hr. The general trends shown in Figure 6 can be summarized as follows: (1) for a given PNA compound the initial rate of transformation increases with increasing initial soil concentration, (2) within the class of polycyclic aromatic compounds, the initial rate of transformation decreases with increasing number of fused benzene rings (or molecular size).

Table XVII. *Kinetic parameters describing rates of degradation of aromatic compounds in soil system.*

PNA	Initial concentration (μg/g soil)	k (day^{-1})	Rate of transformation (μg/g-day)	$t_{1/2}$[a] (days)	Reference
Pyrocatechol	500	3.47	1,735	0.2 m	Medvedev & Davidov (1972)
Phenol	500	0.693	364.5	1.0 m	Medvedev & Davidov (1972)
Phenol	500	0.315	157.5	2.2 l	Medvedev & Davidov (1972)
Fluorene	0.9	0.018	0.016	39 m	Groenewegen & Stolp (1976)
Fluorene	500	0.347	173.3	2 m	Medvedev & Davidov (1972)
Indole	500	0.693	364.5	1.0 m	Medvedev & Davidov (1972)
Indole	500	0.315	157.5	2.2 l	Medvedev & Davidov (1972)
Naphthol	500	0.770	385	0.9 m	Medvedev & Davidov (1972)
Naphthalene	7.0	5.78	40.4	0.12 m	Herbes & Schwall (1978)
Naphthalene	7.0	0.005	0.035	125 l	Herbes & Schwall (1978)
Naphthalene	25,000	0.173	4,331	4 h	Sisler & Zobell (1947)
1,4-Naphthoquinone	500	0.578	288.8	1.2 m	Medvedev & Davidov (1972)
Acenaphthene	500	0.173	86.6	4 m	Medvedev & Davidov (1972)
Acenaphthene	5	2.81	22.6	0.3 m	Medvedev & Davidov (1972)
Anthracene	3.4	0.21	0.714	3.3 l	Gardner *et al.* (1979)
Anthracene	13.7	0.004	0.054	175 m	Gardner *et al.* (1979)
Anthracene	10.3	0.005	0.050	143 m	Gardner *et al.* (1979)
Anthracene	11.4	0.006	0.073	108 m	Gardner *et al.* (1979)
Anthracene	40.0	0.005	0.208	138 m	Gardner *et al.* (1979)
Anthracene	36.4	0.005	0.196	129 m	Gardner *et al.* (1979)
Anthracene	25,000	0.198	4,950	3.5 h	Sisler & Zobell (1947)
Phenanthrene	2.1	0.027	0.056	26 m	Groenewegen & Stolp (1976)
Phenanthrene	25,000	0.277	6,930	2.5 h	Sisler & Zobell (1947)
Carbazole	500	0.067	33	10.5 m	Medvedev & Davidov (1972)
Carbazole	5	0.231	1.16	3 m	Medvedev & Davidov (1972)

Acridine	500	0.075	37.67	9.2	m	Medvedev & Davidov (1972)
Acridine	5	0.281	1.16	3	m	Medvedev & Davidov (1972)
Benz(a)anthracene	0.12	0.046	0.005	15.2	l	Herbes & Schwall (1978)
Benz(a)anthracene	0.12	0.0001	0.00001	6,250	m	Herbes & Schwall (1978)
Benz(a)anthracene	3.5	0.007	0.024	102	m	Groenewegen & Stolp (1976)
Benz(a)anthracene	20.8	0.003	0.062	231	m	Gardner *et al.* (1979)
Benz(a)anthracene	25.8	0.005	0.134	133	m	Gardner *et al.* (1979)
Benz(a)anthracene	17.2	0.008	0.060	199	m	Gardner *et al.* (1979)
Benz(a)anthracene	22.1	0.006	0.130	118	m	Gardner *et al.* (1979)
Benz(a)anthracene	42.6	0.003	0.118	252	m	Gardner *et al.* (1979)
Benz(a)anthracene	72.8	0.004	0.257	196	m	Gardner *et al.* (1979)
Benz(a)anthracene	25,000	0.173	4,331	4	h	Sisler & Zobell (1947)
Fluoranthene	3.9	0.016	0.061	44	m	Groenewegen & Stolp (1976)
Fluoranthene	18.8	0.004	0.072	182	m	Gardner *et al.* (1979)
Fluoranthene	23.0	0.007	0.152	105	m	Gardner *et al.* (1979)
Fluoranthene	16.5	0.005	0.080	143	m	Gardner *et al.* (1979)
Fluoranthene	20.9	0.006	0.125	109	m	Gardner *et al.* (1979)
Fluoranthene	44.5	0.004	0.176	175	m	Gardner *et al.* (1979)
Fluoranthene	72.8	0.005	0.379	133	m	Gardner *et al.* (1979)
Pyrene	3.1	0.020	0.061	35	m	Groenewegen & Stolp (1976)
Pyrene	500	0.067	33	10.5	m	Medvedev & Davidov (1972)
Pyrene	5	0.231	1.16	3	m	Medvedev & Davidov (1972)
Chrysene	4.4	0	0	–		Groenewegen & Stolp (1976)
Chrysene	500	0.067	33	10.5	m	Medvedev & Davidov (1972)
Chrysene	5	0.126	0.63	5.5	m	Medvedev & Davidov (1972)
Benz(a)pyrene	0.048	0.014	0.007	50	l	Herbes & Schwall (1978)
Benz(a)pyrene	0.01	0.001	0.00001	694	l	Herbes & Schwall (1978)
Benz(a)pyrene	3.4	0.012	0.041	57	m	Groenewegen & Stolp (1976)
Benz(a)pyrene	9.5	0.002	0.022	294	m	Gardner *et al.* (1979)
Benz(a)pyrene	12.3	0.005	0.058	147	m	Gardner *et al.* (1979)

Table XVII (*continued*)

PNA	Initial concentration (μg/g soil)	k (day^{-1})	Rate of transformation (μg/g-day)	$t_{1/2}$[a] (days)		Reference
Benz(a)pyrene	7.6	0.003	0.020	264	m	Gardner *et al.* (1979)
Benz(a)pyrene	18.5	0.023	0.312	30	m	Gardner *et al.* (1979)
Benz(a)pyrene	17.0	0.002	0.028	420	m	Gardner *et al.* (1979)
Benz(a)pyrene	32.6	0.004	0.129	175	m	Gardner *et al.* (1979)
Benz(a)pyrene	1.0	0.347	0.347	2	h	Shabad *et al.* (1971)
Benz(a)pyrene	0.515	0.347	0.179	2	h	Shabad *et al.* (1971)
Benz(a)pyrene	0.00135	0.139	0.0002	5	h	Shabad *et al.* (1971)
Benz(a)pyrene	0.0094	0.002	0.00002	406	l	Shabad *et al.* (1971)
Benz(a)pyrene	0.545	0.011	0.006	66	l	Shabad *et al.* (1971)
Benz(a)pyrene	28.5	0.019	0.533	37	l	Shabad *et al.* (1971)
Benz(a)pyrene	29.2	0	0	–		Shabad *et al.* (1971)
Benz(a)pyrene	9,100	0.018	161.7	39	h	Lijinsky & Quastel (1956)
Benz(a)pyrene	19.5	0.099	1.93	7	h	Poglazova *et al.* (1967 b)
Benz(a)pyrene	19.5	0.139	2.70	5	h	Poglazova *et al.* (1967 b)
Benz(a)pyrene	19.5	0.231	4.50	3	h	Poglazova *et al.* (1967 b)
Benz(a)pyrene	130.6	0.173	22.63	4	h	Poglazova *et al.* (1968)
Benz(a)pyrene	130.6	0.116	15.08	6	h	Poglazova *et al.* (1968)
Dibenz(a,h)anthracene	9,700	0.033	320.1	21	h	Lijinsky & Quastel (1956)
Dibenz(a,h)anthracene	25,000	0.039	962.5	18	h	Sisler & Zobell (1947)

[a] l = low temperature range (< 15°C), m = medium temperature range (15-25°C), and h = high temperature range (> 25°C).

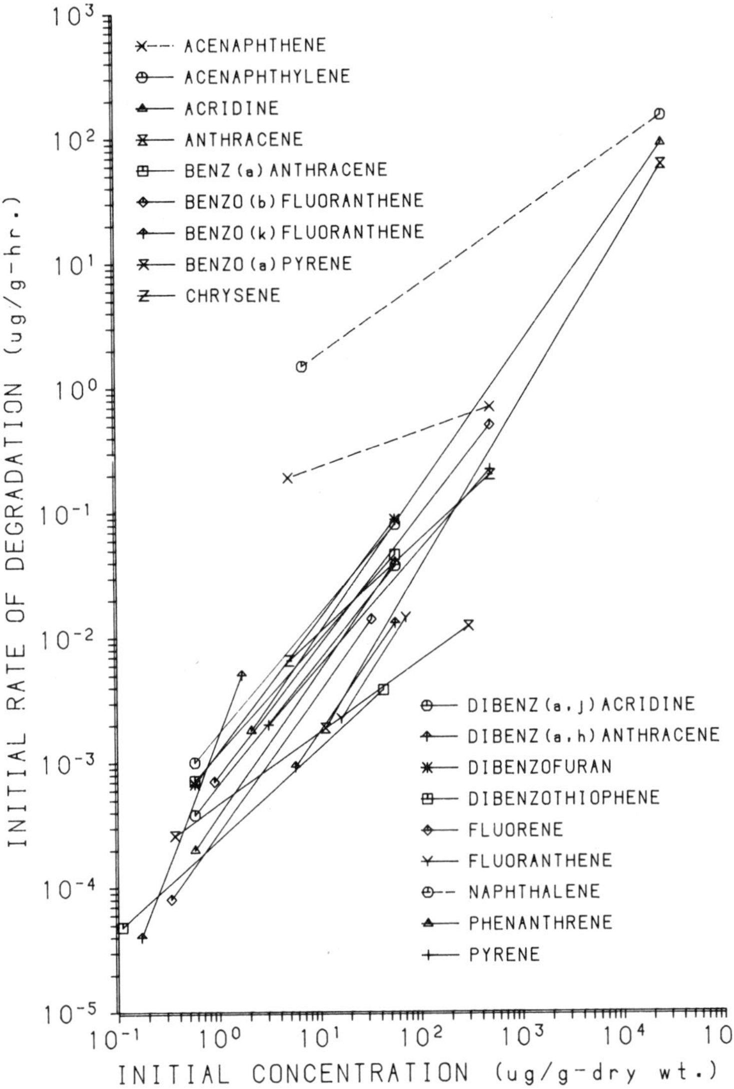

Fig. 6. Rates of transformation on PNA compounds in soil as a function of initial soil concentrations (Sims 1982).

III. Photodecomposition, oxidation, and hydrolysis of PNAs

PNAs, in general, absorb solar radiation strongly and therefore undergo direct photolysis (Radding *et al.* 1976). Table XVIII summarizes photodecomposition information for several PNAs.

PNAs photodecompose readily in the atmosphere by reaction with ozone, other oxidants, nitrogen oxides, and sulfur oxides (NAS 1972). Studies by Southworth (1977), Suess (1972 a), and Andleman and Suess (1971) indicated that PNA photodecomposition in aqueous environments is influenced by water depth, turbidity, solar radiation intensity (seasonal and diurnal fluctuations), temperature, and dissolved oxygen. Reduction of B(a)P under water treatment plant conditions with ultraviolet (UV) radiation amounted to 60 to 80% with 2.5 min of irradiation, and 80 to 90% with 15 min of irradiation (Il'Nitskii *et al.* 1971). The mechanism of photodecomposition for the degradation of PNAs in sediments, sludges, and deposits is believed to be relatively unimportant due to lack of radiation and oxygen beneath surface layers.

The effects of sunlight and UV on photodecomposition of dimethylbenzanthracene (DMBA), B(a)P, and benz(a)anthracene (B(a)A) were investigated by Paalme and Gubergrits (1972). The presence of O_2 had a greater effect than the presence of N_2. For PNAs in general the singlet oxygen is the oxidant and the reaction products include quinones (NAS 1972, Stevens and Algar 1968).

Mailath *et al.* (1974) demonstrated the influence of active surfaces on B(a)P photodecomposition. Sunlight destroyed 65 to 95% in 5 hr with B(a)P on thin silica gel layers. With glass surfaces a 100-fold reduction in sunlight decomposition was observed. Rohrlich and Suckow (1973) obtained 80% degradation of B(a)P adsorbed to silica gel and recovered 13 photodecomposition products, including quinones.

Detoxication of PNAs by irradiation has been investigated to a limited extent. While Suess (1972 b) observed that irradiation products of B(a)P seemed not to be carcinogenic to mice, Graf and Haller (1977) reported that UV degradation products of B(a)P retained 50% of the original toxicity in a microbiological test for carcinogenicity using *Tetrahymena pyriformis*.

The effect of solvents and surfactants on photodecomposition of PNAs has also been investigated (Borneff and Knerr 1959, Suess 1972 a, Paalme *et al.* 1975, and Boyland 1933). Generally, increasing the solubilities of PNAs results in an increase in the extent and rate of photodecomposition. The presence of phenols in aqueous PNA solutions has been shown to increase the rate of photochemical oxidation (Karu *et al.* 1973), and to inhibit the extent of photochemical oxidation (Gubergrits *et al.* 1972).

Photodecomposition of PNAs has also been observed to be influenced by the presence of other PNAs. B(a)P photolysis was increased in the presence of B(e)P (Paalme and Gubergrits 1974), DMBA (Paalme *et al.* 1976), and 3-methylcholranthrene (Paalme and Gubergrits 1976). However, in the same study, pyrene decreased B(a)P photolysis.

While singlet oxygen is considered to be the major oxidant species generated by the direct photolysis of PNAs, other principal oxidizing species in natural aqueous systems include alkylperoxy ($RO_2\cdot$) and hydroperoxy ($HO_2\cdot$) radicals generated by photolytic cleavage of trace carbonyl compounds or from enzymatic sources (Versar 1979). Free radical oxidation of PNAs are summarized by

Radding *et al.* (1976). Half-lives for the reaction of $RO_2\cdot$ radical with anthracene, B(a)P, and perylene have been calculated to be 1,600, 9,900, and 1,600 days, respectively, for a 10^{-10} molar steady-state concentration for the $RO_2\cdot$ radical for average daily illumination in natural waters (Radding *et al.* 1976).

Hydrolysis is not considered to be an important mechanism for degradation of PNAs in the environment. Polycyclic aromatic hydrocarbons do not contain groups amenable to hydrolysis (Radding *et al.* 1976).

Photodecomposition, oxidation, and hydrolysis of PNA compounds are not considered as significant pathways for PNA degradation in soil environments.

IV. Volatilization of PNAs

Volatilization half-lives for most PNAs are not known. Table XIX summarizes available information for several PNAs. The rate of volatilization from aqueous environments is influenced by properties of the compound (Henry's law constant) and the environment (wind, water turbulence, and temperature) (Southworth 1977).

Volatilization is considered more important under environmental conditions characterized by shallow, clear, agitated natural waters. Methods for predicting volatilization rates from natural aqueous systems using theoretical considerations (Mackay and Wolkoff 1973, Mackay and Leinonen 1975) require substantial measured physical data and are not considered practical for many conditions (Versar 1979). Southworth (1977) employed a simple first-order exponential decay equation and considered Henry's Law constants, gas phase exchange, and liquid phase exchange for the environmental conditions listed above. He concluded that volatilization of polycyclic aromatic hydrocarbons with four or more rings is expected to be insignificant in any sizeable water body due to the low Henry's Law constants. Although volatilization of some high molecular weight, low-water-soluble organics has been shown to be significant due to high activity coefficients (Mackay and Wolkoff 1973), volatilization is not considered to be an important transport and removal process for PNAs (Versar 1979, Southworth 1977) (refer to Table II for vapor pressure for PNAs).

Smith *et al.* (1978) state that when PNAs are sorbed onto sediments, volatilization of the sorbed PNAs is presumed to be very slow. The relatively high log octanol:water partition coefficients of most PNAs indicate that an adsorption mechanism is much more important in determining the fate of PNAs in terrestrial systems than volatilization.

V. Soil adsorption, leaching, and run-off

Adsorption and leaching characteristics of PNAs in soils and sediments are needed to predict their movement and fate in terrestrial and aquatic environments. Adsorption of nonionic, nonpolar chemicals by solid surfaces and movement through porous media depends upon chemical properties of the compound

Table XVIII. *Photodecomposition of PNAs.*

PNA compound	Comments and $t_{1/2}$	Reference
Naphthalene	absorbs solar radiation strongly	Versar (1979)
Anthracene	$t_{1/2}$ = 35 min, midday sunlight, midsummer, 35°N latitude	Southworth (1977)
Fluorene	absorbs solar radiation strongly	Versar (1979)
Acenaphthylene	absorbs solar radiation strongly	Versar (1979)
Acenaphthene	absorbs solar radiation strongly	Versar (1979)
Phenanthrene	absorbs solar radiation strongly	Versar (1979)
Benz(a)anthracene	$t_{1/2}$ = 4.8 hr, winter, 35°N latitude	Southworth (1977)
	UV degradation creates quinones	McGinnes & Snoeyink (1974)
Chrysene	absorbs solar radiation strongly	Versar (1979)
Pyrene	absorbs solar radiation strongly	Versar (1979)
Fluoranthene	absorbs solar radiation strongly	Versar (1979)
Benz(a)pyrene	$t_{1/2}$ = 1 hr, midday, winter	Smith *et al.* (1978)
	adsorption to kaolinite clay inhibits photolysis	Lu *et al.* (1977)
	$t_{1/2}$ = 11 hr (21°C, O_2) with fluorescent light on B(a)P adsorbed on calcite and suspended in water	Suess (1972 a)
	$t_{1/2}$ = 20 hr with UV radiation	Rondia & Epstein (1968)
	surfactants increase photolysis rates	Paalme *et al.* (1975)
	$t_{1/2}$ = 5.3 hr in presence of UV and without O_3	Lane & Katz (1977)
	$t_{1/2}$ = 0.2 hr in presence of both UV and O_3	Lane & Katz (1977)
Benzo(k)fluoranthene	absorbs solar radiation strongly	Versar (1979)
Dibenz(a,h)anthracene	absorbs solar radiation strongly	Versar (1979)
7,12-Dimethylbenzanthracene	absorbs solar radiation strongly; main product is	Versar (1979)
	epoxide; rate is function of solvent, wave length and intensity	Lopp *et al.* (1976)

3-Methylcholanthrene	rate catalyzed by dimethyl sulfoxide solvent	Dao *et al.* (1971)
Benzo(k)fluoranthene	absorbs solar radiation strongly	Versar (1979)
Benzo(ghi)perylene	absorbs solar radiation strongly	Versar (1979)
Indeno(1,2,3-cd)pyrene	absorbs solar radiation strongly	Versar (1979)

Table XIX. *Volatilization of PNAs.*

PNA compound	Comments and $t_{1/2}$	Reference
Naphthalene	Vapor pressure high (important process)	Versar (1979)
Anthracene	$t_{1/2}$ = 300 hr under quiescent conditions in water 1 m deep	Southworth (1977)
	$t_{1/2}$ = 18 hr under agitation	Southworth (1977)
Fluorene	not important process	Versar (1979)
Acenaphthalene	not important process	Versar (1979)
Acenaphthene	not important process	Versar (1979)
Phenanthrene	not important process	Versar (1979)
Benz(a)anthracene	$t_{1/2}$ = 90 hr; rate constant 0.00789 hr^{-1}	Smith *et al.* (1978)
Chrysene	not important process	Versar (1979)
Pyrene	not important process	Versar (1979)
Fluoranthene	not important process	Versar (1979)
Benz(a)pyrene	$t_{1/2}$ = 22 hr with agitation; rate constant 0.030 hr^{-1} adsorption reduces volatilization significantly	Smith *et al.* (1978)
Benzo(f)fluoranthene	not important process	Versar (1979)
Dibenz(a,h)anthracene	not important process	Versar (1979)
Benzo(k)fluoranthene	not important process	Versar (1979)
Benzo(ghi)perylene	not important process	Versar (1979)
Indeno(1,2,3-cd)pyrene	not important process	Versar (1979)

Table XX. *Adsorption of PNAs by activated carbon and by soils and sediments.*

PNA	Log K_p [a]	Adsorbent: Activated carbon [b]		Adsorbent: Soils and sediments	
		K	1/n	K	K_{oc}
Benzene	–	1.0	1.60	0.13 [e]	83 [d]
Toluene	–	26.1	0.44	–	–
Quinoline	–	–	–	10.9 [d]	–
Naphthalene	3.37	132	0.42	–	1,300 [d]
2-Methylnaphthalene	–	–	–	–	8,500 [d]
Benzo(b)thiophene	–	–	–	50	–
Anthracene	4.45	376	0.70	0.01-0.05 [c]	26,000 [d]
Benzo(f)quinoline	–	–	–	1,313	–
Acenaphthylene	4.07	115	0.37	–	–
Acenaphthene	4.33	190	0.36	–	–
Phenanthrene	4.46	215	0.44	–	23,000 [d]
Fluorene	4.18	330	0.28	–	–
9H-carbazole	–	–	–	175 [d]	–
Fluoranthene	5.33	664	0.61	–	–
Tetracene	–	–	–	–	650,000 [d]
Pyrene	5.32	–	–	–	84,000 [d]
Benz(a)anthracene	5.61	–	–	26,200 [d]	–
Dibenzothiopene	–	–	–	137 [g]	10,240 (mean) [g]
3-Methylcholanthrene	6.97	–	–	18,196 (mean) [f]	1,789,000 (mean) [f]
Benzo(b)fluoranthene	6.57	57	0.37	–	–
Benzo(k)fluoranthene	6.84	181	0.57	–	–

Benz(a)pyrene	6.04	33.6	0.44	76,000[d]	4,510,651(mean)[d]
Dibenz(a,h)anthracene	5.97	69.3	0.75	22,774(mean)[f]	2,029,000(mean)[f]
Benzo(ghi)perylene	7.23	10.7	0.37	–	–
7H-dibenzocarbazole	–	–	–	26,233(mean)[d]	–

[a] Versar (1979).
[b] Dobbs & Cohen (1980).
[c] Herbes (1977).
[d] Reinbold *et al.* (1978).
[e] McBride *et al.* (1975).
[f] Means *et al.* (1980).
[g] Hassett *et al.* (1980).

(adsorbate) and the characteristics of the solid surface (adsorbent). Thus organic matter presence is highly correlated with adsorption. Water soluble compounds tend to migrate in soil and aquatic systems and are not significantly adsorbed. However, more hydrophobic compounds become readily adsorbed to lipophilic organic matter in soils and, as a result, become relatively non-mobile.

Reinbold *et al.* (1978) summarized the adsorption of energy-related pollutants in the environment. A review of the literature on the adsorption of pollutants on sediments and soils revealed that, except for pesticides, most sorption information for organic chemicals was obtained using adsorbents other than soils, sediments, and clay minerals. Adsorption constants were most often derived by use of Freundlich or Langmuir adsorption equations.

The mechanism of adsorption is used to explain the bioavailability of organic chemicals between different soils or different classes of chemicals (Bailey and White 1964). Major factors influencing adsorption in soil systems include soil type, physicochemical properties of the chemicals, soil moisture, temperature, presence of competing molecules, solvents, and pH.

Table XX summarizes adsorption information for several PNAs. Included are octanol-water partition coefficients, Freundlich activated carbon isotherm results (adsorption capacity (K) and adsorption intensity (1/n)), and soil/sediment adsorption results using the Freundlich equation (adsorption capacity (K) and an adjustment of capacity based on organic carbon content (K_{oc})). The Freundlich equation can be expressed as

$$X/M = K\,C_f^{1/n}$$

where

X/M = unit adsorbate removed from solution/unit adsorbent,
C_f = amount of adsorbate remaining in the treated water at equilibrium,
K = a coefficient describing the adsorption capacity, and
$1/n$ = indication of adsorption intensity.

When adsorption is normalized with respect to organic carbon (OC) content of the adsorbent, the form of the equation used is

$$K_{oc} = \frac{(K)\,(100)}{\%\,OC}$$

McBride *et al.* (1975) investigated the adsorption of benzene on smectite (Wyoming montmorillonite) as a function of organophilic cation and water content. Results are presented in Table XXI. Competition between water and the aromatic molecules has the most obvious effect on adsorption of benzene. Langmuir adsorption isotherm results indicated a maximum benzene adsorption with TMA+-montmorillonite of 13.2 g benzene/100 g adsorbent.

The sorption of pyrene, 7,12-dimethylbenzanthracene, 3-methylcholanthrene, and dibenz(a,h)anthracene on 14 soil/sediment samples was investigated by Means *et al.* (1980). Results for soil/sediment characteristics and PNA adsorption are summarized in Table XXII. Freundlich equilibrium constants (K_d) were obtained and since the isotherms were linear, the data were fitted to a simplified equation:

$$C_s = K\,C_w$$

The resulting partition constants (K) are presented in Table XXII. The sorption constants (K), when normalized to organic carbon content of the adsorbent (K_{oc}), were predictive of the octanol-water partition coefficient for the PNAs listed:

$$\log K_{oc} = \log K_{ow} - 0.317$$

where K_{ow} indicates the octanol-water partition coefficient.

Although many soil/sediment characteristics were determined, including pH, CEC, total N, organic carbon, and sand, silt, and clay content, the only significant relationship found was that between K and percent organic carbon of the adsorbent (Table XXII). K_{oc} has been shown to be a unique constant for non-polar compounds (Karickhoff *et al.* 1979, Means *et al.* 1979, Means *et al.* 1980).

The order of increasing adsorption to the soils/sediments is: pyrene < 7,12-dimethylbenzanthracene < 3-methylcholanthrene < dibenz(a,h)anthracene. The increase in sorption of these PNAs appears to be more closely related to increases in effective chain length of the molecule than to increases in mass alone.

The important role of organic matter versus mineral particulates was emphasized by Herbes (1977). Adsorption of 0.02 μg/L anthracene by montmorillonite concentrations of 10 to 100 mg/L ranged from one to 5%. Adsorption of anthracene by autoclaved yeast cells under identical conditions results in 10 to 50% uptake. The heat of adsorption to the organic matter (yeast cells) was 5.2 kcal/mole indicating a physical adsorption process.

The influence of soil type on adsorption of pyrene was investigated by Karickhoff *et al.* (1979). They calculated the partition coefficient (K_p) and the correction for organic carbon (K_{oc}) for each soil type. Results are presented in Table XXIII. The significant role of the organic carbon in soil in PNA adsorption is obvious.

Table XXI. *Benzene adsorption on Wyoming montmorillonite* (McBride *et al.* 1975).

Exchange form of Smectite	Benzene adsorbed (g/100 g) From vapor (no water)	Benzene adsorbed (g/100 g) From aq solution (350 mg/L)
Cu^{2+}	18.8	0.6
Tetramethylammonium (TMA^+)	22.6	7.2
Tetraethylammonium (TEA^+)	14.2	1.8
Tetrapropylammonium (TPA^+)	13.8	0.0

Table XXII. *Soil and sediment properties and sorption constants for PNAs* (Means *et al.* 1980).

Soil and sediment properties				PNA adsorption characteristics							
				Pyrene		DMBA		3-MC		DB(a,h)A	
Type	CEC (meq/100g)	Organic carbon (%)	Clay (%)	K_p	K_{oc}	K_p	K_{oc}	K_p	K_{oc}	K_p	K_{oc}
B2	3.72	1.21	18.6	760	62,860	2,371	195,998	15,140	1,251,210	20,461	1,690,971
4	23.72	2.07	55.2	1,065	51,469	2,648	127,812	30,085	1,453,404	34,929	1,687,404
5	19.00	2.28	31.0	1,155	50,650	5,210	228,499	8,273	362,845	18,361	805,292
6	33.01	0.72	68.6	814	85,256	1,346	186,986	15,820	2,197,250	19,882	2,622,453
8	3.72	0.15	8.8	101	67,467	611	407,496	2,257	1,504,538	1,759	1,172,847
9[a]	12.40	0.11	17.4	71	64,706	1,028	934,225	2,694	2,449,190	2,506	2,277,875
14[a]	18.86	0.48	63.6	277	57,763	562	117,161	30,627	6,380,703	14,497	3,020,262
15	11.30	0.95	35.7	788	82,421	3,742	393,907	23,080	2,429,456	25,302	2,663,317
18	15.48	0.66	39.5	504	76,816	1,895	287,196	20,642	3,127,521	20,192	3,059,425
20[a]	8.50	1.80	28.0	723	59,946	1,617	124,347	16,231	1,248,534	7,345	565,014
21	8.33	1.88	7.1	1,119	59,515	5,576	296,580	24,506	1,303,532	55,697	2,962,603
22	8.58	1.67	21.2	806	48,236	2,679	160,801	20,972	1,255,821	39,809	2,383,765
23	31.15	2.38	69.1	1,043	43,897	6,777	284,743	17,127	719,693	19,254	808,991
26	20.86	1.48	42.9	994	67,189	8,740	252,735	37,364	2,524,581	30,840	2,691,870

[a] Soils; all others are sediments.

Table XXIII. *Adsorption of pyrene and dibenzothiophene to soils.*

Soil		Pyrene adsorption characteristics[a]	
Type	Organic carbon (%)	K_p	K_{oc}
Hickory Hill			
sand	0.18	42	32,000
coarse silt	3.27	3,000	92,000
medium silt	1.98	2,500	1,300,000
fine silt	1.34	1,500	1,100,000
clay	1.20	1,400	1,200,000
Sample		Dibenzothiophene adsorption characteristics[b]	
		K_p	K_{oc}
B2		117.5	9,711
4		180.6	8,725
5		167.1	7,329
6		60.8	8,444
8		9.4	6,267
9		5.8	5,273
14		49.7	10,354
15		179.9	18,937
18		65.1	9,864
20		101.4	7,800
21		276.0	14,681
22		176.3	10,557
23		388.6	16,328
26		134.5	9,088

[a] Karickhoff *et al.* (1979).
[b] Hassett *et al.* (1980).

The influence of soil characteristics on adsorption of dibenzothiophene was investigated by Hassett *et al.* (1980). They calculated the partition coefficient (K_p) and the correction for organic matter (K_{oc}) for each sample. Results are presented in Table XXIII. Again the significant role of the organic carbon in soil in PNA adsorption is obvious.

VI. Plant uptake, effects, and metabolism of PNAs

a) Natural background concentration of PNAs in vegetation

PNAs are ubiquitous constituents of crops, plants, and algae in the natural environment including urban and pristine areas. Natural background levels of PNAs reported for various plants and crops are summarized in Table XXIV. In

Table XXIV. *Natural background levels of PNAs in plants/crops.*

PNA	Plant/crop	Concentration (μg/kg, dry wt)	Reference
Anthracene	Post oak (leaves)	70	Hancock *et al.* (1970)
Anthracene	Little Bluestem (leaves)	50	Hancock *et al.* (1970)
Fluoranthene	Little Bluestem (leaves)	33	Hancock *et al.* (1970)
Fluoranthene	Post oak (leaves)	62	Hancock *et al.* (1970)
Benz(a)anthracene	Little Bluestem (leaves)	22	Hancock *et al.* (1970)
Benz(a)anthracene	Post oak (leaves)	60	Hancock *et al.* (1970)
Pyrene	Little Bluestem (leaves)	58	Hancock *et al.* (1970)
Pyrene	Post oak (leaves)	88	Hancock *et al.* (1970)
Benz(a)pyrene	Little Bluestem (leaves)	30	Hancock *et al.* (1970)
Benz(a)pyrene	Post oak (leaves)	40	Hancock *et al.* (1970)
Benz(a)pyrene	Lettuce	12	Graf & Diehl (1966)
Benz(a)pyrene	Spinach	20	Graf & Diehl (1966)
Benz(a)pyrene	Endive	50	Graf & Diehl (1966)
Benz(a)pyrene	Potato peeling	0.36	Shabad & Cohan (1972)
Benz(a)pyrene	Potato tuber	0.09	Shabad & Cohan (1972)
Benz(a)pyrene	Spring wheat seed	0.29	Shabad & Cohan (1972)
Benz(a)pyrene	Spring wheat stem	27.0	Shabad & Cohan (1972)
Benz(a)pyrene	Winter rye seed	0.35	Shabad & Cohan (1972)
Benz(a)pyrene	Coffee green beans	0.60	Il'Nitskii & Cohan (1972)
Benz(a)pyrene	Young wheat plants	48-66	Siddiqi & Wagner (1972)
Benz(a)pyrene	Carrots	1.65-6.0	Il'Nitskii *et al.* (1974)
Benz(a)pyrene	Beets	0.6-2.0	Il'Nitskii *et al.* (1974)
Benz(a)pyrene	Cabbage	0.8-1.2	Il'Nitskii *et al.* (1974)
Benz(a)pyrene	Carrots	0.08-0.14	Siegfried (1975)
Benz(a)pyrene	Head lettuce	0.2-1.3	Siegfried (1975)
Benz(a)pyrene	Japanese radish root	0.01	Shiraishi & Takabatake (1974)

Benz(a)pyrene	Spinach	0.26-3.3	Shiraishi & Takabatake (1974)
Benz(a)pyrene	Chrysanthemum	1.20	Shiraishi & Takabatake (1974)
Benz(a)pyrene	Chinese cabbage	0.05	Shiraishi & Takabatake (1974)
Benz(a)pyrene	Potato	0.01	Shiraishi & Takabatake (1974)
Benz(a)pyrene	Carrot	0.02	Shiraishi & Takabatake (1974)
Benz(a)pyrene	Onion	0.01	Shiraishi & Takabatake (1974)
Benz(a)pyrene	Sunflower oil, cottonseed oil, peanut oil, rapeseed oil	1-5	Grigorenko *et al.* (1970)
Benz(a)pyrene	Apple	0.02	Shiraishi (1975)
Benz(a)pyrene	Banana	0.02	Shiraishi (1975)
Benz(a)pyrene	Banana peel	0.03	Shiraishi (1975)
Benz(a)pyrene	Grape	0.02	Shiraishi (1975)
Benz(a)pyrene	Persimmon	0.02	Shiraishi (1975)
Benz(a)pyrene	Pineapple	0.02	Shiraishi (1975)
Benz(a)pyrene	Plum	0.04	Shiraishi (1975)
Benz(a)pyrene	Mandarin orange	0.03	Shiraishi (1975)
Benz(a)pyrene	Orange peel	0.15	Shiraishi (1975)

general leaves exhibit the largest PNA concentrations (22 to 88 ppb), and underground vegetables including potatoes, carrots, onions, and radishes exhibit the lowest PNA concentrations (0.01 to 6.0 ppb). The distribution of PNAs among various categories of vegetation is approximately: tree leaves (22 to 88 ppb) > cereals (48 to 66 ppb) > leafy vegetables (0.05 to 50 ppb) > underground vegetables (0.01 to 6.0 ppb) > fruits (0.02 to 0.04 ppb).

While the presence of background levels of PNAs in vegetation is readily demonstrated (Table XXIV), the source and the relative contributions of different possible sources of PNAs is the subject of much controversy and debate. Atmospheric deposition of PNAs produced as a result of industrial processes is considered the source of benz(a)pyrene (B(a)P) in soils and in vegetation near air fields (Smirnov 1970) and near urban areas (Kolar *et al.* 1975). Aboveground plant parts contained higher levels of B(a)P than parts beneath the ground. Also the amount of B(a)P was proportional to the exposed plant surface area. These studies indicate that atmospheric deposition is the source of PNAs in vegetation rather than uptake from polluted soils. The degree of PNA contamination of vegetables is often explained by surface area/kg, location, and exposure time (Grimmer and Hildebrand 1965).

Other researchers (Hancock *et al.* 1970, Graf and Diehl 1966, and Borneff *et al.* 1968 a and b) have demonstrated that both higher and lower plants exhibit background levels of PNAs independent of the area in which they are grown. The presence of a 'base level' of PNAs in plants is attributed to biochemical synthesis by the plant.

In addition to anthropogenic activities resulting in PNA deposition on plants, and biochemical synthesis by plants, a third possible source of PNAs found in vegetation is plant-uptake from soil. Uptake of PNAs from aqueous and soil media and from compost has been investigated by Kolar *et al.* (1975), Shabad and Cohan (1972), Shcherbak (1969), Wagner and Siddiqi (1970), Borneff *et al.* (1973), Siegfried (1975), and Graf and Nowak (1966), among others. Unfortunately, the major emphasis of research has been devoted to the study of B(a)P behavior, which does not always correlate with the behavior of other PNAs.

The health implications of ingestion of plants with background PNA levels are not known. Hakama and Saxen (1967) and others (NAS 1972) have demonstrated correlations of human stomach cancer and high cereal consumption and ingestion of plants containing higher levels of PNAs. Wynder *et al.* (1963) attributed as significant the role of starchy foods with concomitant low intake of fresh fruits as the cause of cancer among vegetarians. However, because plant parts containing starch contain only one to 10% of the amount of PNA in leaves and green plant parts, the role and significance of plant or plant parts in the human diet is not known.

PNAs in soils and in wheat plant parts are summarized in Table XXV. With B(a)P, the crop/soil ratio for spring and summer wheat was lowest in seeds and much higher in stems and straw. Also the concentration of B(a)P in the plant parts was not a function of the soil concentration, as shown in Table XXV.

Table XXV. *PNAs in soils and in wheat.*

PNA	Wheat Type	Wheat Concentration (μg/kg, dry wt)	Soil Concentration (μg/kg, dry wt)	Crop/Soil	Reference
Benz(a)pyrene	Spring wheat-seed	0.29	1.6	0.18	Shabad & Cohan (1972)
Benz(a)pyrene	Spring wheat-seed	0.38	170	0.002	Shabad & Cohan (1972)
Benz(a)pyrene	Spring wheat-seed	0.84	1.6	0.53	Shabad & Cohan (1972)
Benz(a)pyrene	Spring wheat-seed	0.72	170	0.004	Shabad & Cohan (1972)
Benz(a)pyrene	Spring wheat-stem	4.52	1.6	2.9	Shabad & Cohan (1972)
Benz(a)pyrene	Spring wheat-stem	3.60	170	0.021	Shabad & Cohan (1972)
Benz(a)pyrene	Spring wheat-straw	27.0	1.6	17.0	Shabad & Cohan (1972)
Benz(a)pyrene	Spring wheat-straw	26.7	170	0.16	Shabad & Cohan (1972)
Benz(a)pyrene	Summer wheat-seed	3.0	2.0	1.5	Wagner & Siddiqi (1970)
Benz(a)pyrene	Summer wheat-seed	3.0	1,300	0.002	Wagner & Siddiqi (1970)
Benz(a)pyrene	Summer wheat-stem	15.0	2.0	7.5	Wagner & Siddiqi (1970)
Benz(a)pyrene	Summer wheat-stem	25.0	1,300	0.020	Wagner & Siddiqi (1970)
Benz(a)pyrene	Summer wheat-straw	7.0	2.0	3.5	Wagner & Siddiqi (1970)
Benz(a)pyrene	Summer wheat-straw	8.0	1,300	0.01	Wagner & Siddiqi (1970)
3,4-Benzfluoranthene	Summer wheat-seed	76	4.4	17.0	Wagner & Siddiqi (1970)
3,4-Benzfluoranthene	Summer wheat-seed	98	1,300	0.08	Wagner & Siddiqi (1970)
3,4-Benzfluoranthene	Summer wheat-stem	100	4.4	2.30	Wagner & Siddiqi (1970)
3,4-Benzfluoranthene	Summer wheat-stem	223	1,300	0.18	Wagner & Siddiqi (1970)
3,4-Benzfluoranthene	Summer wheat-straw	104	4.4	24.0	Wagner & Siddiqi (1970)
3,4-Benzfluoranthene	Summer wheat-straw	220	1,300	0.18	Wagner & Siddiqi (1970)

When the soil concentration of B(a)P was increased, the plant concentration remained approximately the same. However, biomagnification of B(a)P for seed, stem and straw is demonstrated. With 3,4-benzfluoranthene, the pattern was not the same. Plant concentrations were generally much higher, with smaller differences among the plant parts. Also an increase in plant 3,4-benzfluoranthene content correlated with an increase in the soil PNA content. Biomagnification of 3,4-benzfluoranthene is demonstrated. Thus the two PNAs exhibited different behavior under similar soil and crop conditions. Biomagnification, however, is evident with both PNAs.

Soil and plant levels of PNAs as influenced by fertilizer type are presented in Table XXVI. For horseradish the crop/soil ratio is consistently lower for the root center and higher for the root skin, regardless of the method of fertilization. Interestingly, the crop/soil ratios tend to be as high or higher for non-fertilized and mineral fertilized soils as for compost and manure fertilized soils. With carrots, B(a)P content in the plant did not correlate with B(a)P content in the soil. Biomagnification is not indicated for any of the crops, as plant PNA content varied from 0.04% to 78% of the soil content.

Borneff *et al.* (1973) concluded that B(a)P content in food is small and is unrelated to compost origin or B(a)P content of the compost. PNA content in industrially prepared compost is given by Wagner and Vonderheid (1978). Only in the outer layer of plant roots does the B(a)P concentration correlate with the soil concentration (Borneff *et al.* 1973). Because of the small background and concentration in plants, the amount ingested with edible roots grown on compost-amended soil may not enhance the human health hazard.

Much information has been collected which indicates that many higher plants may not take up PNAs. Blum and Swarbrick (1977) studied uptake of radiolabeled ^{14}C-B(a)P by plants grown hydroponically in nutrient solutions containing 8.0 ppb B(a)P. Crops investigated included green beans, cantalopes, and cotton seeds. Results indicated that no (<3 ppb) B(a)P was translocated or accumulated in any of the plants or plant tissues, including leaf and stem tissues for green beans.

b) Plant response to PNA additions

PNAs have been demonstrated to act like plant hormones, stimulating growth and yield of higher and lower plants. Table XXVII presents the results of the effect of B(a)P on algal cell size and summer rye kernel yield. Graf (1965) demonstrated the growth-promoting effects of PNAs with higher plants. Daily watering with B(a)P (10 μg/L) effected an increase in yield of kohlrabi (20%), cauliflower (20%), rye (300%), and tobacco (100%).

Graf (1965) also demonstrated that the growth promoting effect was proportional to the carcinogenic potential. Dibenz(a,h)anthracene which is a more potent carcinogen than benz(a)anthracene had a stronger growth-promoting effect on tobacco.

Table XXVI. *Average soil and plant levels of PNAs*[a] *as influenced by fertilizer type.*

Crop	Treatment														
	Fresh compost			Piled compost			Stable manure			Mineral fertilizer			Non-fertilized		
	Soil	Crop	C/S[c]	Soil	Crop	C/S[c]	Soil	Crop	C/S[c]	Soil	Crop	C/S[c]	Soil	Crop	C/S[c]
	(μg/kg, dry wt)			(μg/kg, dry wt)			(μg/kg, dry wt)			(μg/kg, dry wt)			(μg/kg, dry wt)		
Horseradish															
root center	537	35.9	0.07	258	16.4	0.06	109.2	14.6	0.13	74.4	17.0	0.23	77.2	15.7	0.20
root skin	537	86.4	0.16	258	71.7	0.28	109.2	51.6	0.47	74.4	57.8	0.78	77.2	42.9	0.56
Red radish															
roots	482	70.4	0.15	221	55.8	0.25	150.2	37.9	0.25	66.3	19.1	0.29	83.9	19.9	0.23
Carrots[b]	85	0.11	0.0013												
Carrots[b]	130	0.16	0.0012												
Carrots[b]	185	0.08	0.004												

[a] PNAs include: fluoranthene, 3,4-benzofluoranthene, 3,4-benzpyrene, 1,12-benzperylene, 11,12-benzfluoranthene, Indeno(1,2,3-cd)pyrene (Borneff *et al.* 1973).

[b] For carrots only 3,4-benzpyrene monitored (Siegfried 1975).

[c] C/S = Crop/Soil ratio.

Table XXVII. *Effect of benz(a)pyrene on algal cell size and crop yield* (Graf and Nowak 1966).

PNA	Algae (*Chlorella* sp.) Size ratio (μ) (Experimental/control)	Algae (*Scenedesmus*) Size ratio (μ) (Experimental/control)	Algae (*Ankistradesmus*) Size ratio (μ) (Experimental/control)	Summer rye kernels Yield ratio (g) (Experimental/control)
Fluoranthene	1.08	1.20	1.21	0.28
1,21-Benzperylene	1.00	1.20	1.26	1.16
3,4-Benzfluoranthene	1.05	3.71	1.47	0.39
Indeno(1,2,3-cd)pyrene	1.11	1.14	1.32	1.63
1,2-Benzanthracene	1.16	1.19	0.97	1.08
3,4-Benzpyrene	1.19	2.07	1.16	1.35

Table XXVIII. *Effect of 3,4-benzfluoranthene in soils with wheat and barley* (Wagner and Wagner-Hering 1971).

3,4-Benzfluoranthene (μg/kg, dry wt) in soil	Plant				
	Wheat			Barley	
	(Yield, dry wt-g)	Concentration (μg/kg, dry wt) Crop	Soil	(yield, dry wt-g) Root	Straw
Control (2.0)	20.8	100	50	10.4	36.8
130	21.3	237	1.83		
1300	20.8	223	0.18		
Control (2.0)	44.9	119	59.4		
3100	38.5	273	0.087		
6300	39.3	483	0.077		
9400				16.9	33.5
19000				12.4	34.3

The 'normal' presence of PNAs in plants and the stimulation of growth upon addition of PNAs to plants led Graf and Nowak (1966) and Graf (1965) to suggest that PNAs may be plant growth promoters (auxins) or their precursors.

While PNAs act to stimulate plant growth, they may not accumulate in the plant. Graf (1965) observed that B(a)P, supplied to tobacco through the roots, did not increase in concentration in the plant relative to the dry weight as compared to control plants. Thus, the B(a)P supplied to the plants is consumed during growing, with no net accumulation in the plant.

Tumor-like growth has also been reported for other plants. Ishio *et al.* (1972) reported neoplastic growth on marine algae. Mushrooms have also exhibited cancer-like growth in the presence of PNAs (NAS 1972).

Other researchers (Wagner and Wagner-Hering 1971) actually obtained growth depression and retardation with PNAs and higher plants. 3,4-Benzfluoranthanthene at 6200 μg/kg effected a reduced stem length of wheat (78 vs. 71 cm). Table XXVIII lists results of the effects of 3,4-benzfluoranthene on wheat and barley. While reduction in the yield of wheat is small, even at high 3,4-benzfluoranthene concentrations, there was an apparent uptake of the PNA which was related to the soil concentration. Apparently, wheat may accumulate 3,4-benzfluoranthene by a factor of five and not show significant yield reductions. At higher soil concentrations, 3,4-benzfluoranthene reduces root weight of barley plants by almost 50%, while reduction in straw yield is minimal. A selective effect of the PNA in retarding plant parts is suggested here.

c) Plant biosynthesis of PNAs

There is strong evidence for plant biosynthesis of PNAs. Borneff *et al.* (1968 a) demonstrated the biosynthesis of seven PNAs by algal cells using ^{14}C acetate as a sole carbon source. Results are presented in Table XXIX. The PNA concentrations are typical of background levels in plants. Graf and Diehl (1966) demonstrated an apparent synthesis of B(a)P by growing seeds of rye, lentils, wheat and hyacinths. Table XXX summarizes the results. These studies were conducted under controlled conditions to prevent the introduction of exogenous PNAs.

Table XXIX. *PNAs content in algae with* ^{14}C*-acetate as sole carbon source* (Borneff *et al.* 1968 b).

PNA	Concentration (μg/kg, dry wt)
Fluoranthene	54.62
Benz(a)anthracene	5.73
Benzo(b)fluoranthene	3.70
Benz(a)pyrene	0.70
Benzo(ghi)perylene	2.03
Benzo(k)fluoranthene	1.32
Indeno(1,2,3-cd)pyrene	1.50

Table XXX. *Benz(a)pyrene levels in higher plant seeds and seedlings grown hydroponically* (Graf and Diehl 1966).

Crop (seeds)	Concentration (μg/kg, dry wt)	Crop (seedlings)	Concentration (μg/kg, dry wt)	Seedlings/ seed ratio
Rye kernels	2.0	rye seedlings	18.0	9
Rye kernels	0.5	rye seedlings	38.0	76
Lentils	0.1	lentil seedlings	24.0	240
Wheat kernels	0.5	wheat seedlings	8.0	16
Hyacinth bulbs	3.0	hyacinths	10.0	3

Table XXXI. *PNA levels in young and old leaves of higher plants (μg/kg, dry wt)* (Graf and Diehl 1966).

	Plant								
	Beech leaf			Oak leaf			Tobacco leaf		
PNA	Green	Yellow	Y/G[a]	Green	Yellow	Y/G[a]	Green	Yellow	Y/G[a]
Fluoranthene	310	1200	4.0	126	666	5.3	376	626	1.7
Benzo(b)fluoranthene	62	374	6.0	62	292	4.7	32	126	3.9
Indeno(1,2,3-cd)pyrene	38	234	6.0	26	150	5.8	18	38	2.1
Benz(a)anthracene	26	166	6.4	18	100	5.6	12	44	3.7
Benzo(ghi)perylene	38	142	3.7	16	74	4.6	18	44	2.4
Benz(a)pyrene	16	66	4.1	8	42	5.3	12	38	3.2
Average			5.0			5.2			2.8

[a] Y/G = Yellow/Green ratio.

It was also reported by Graf and Diehl (1966) that the content of beech, oak, and tobacco leaves increased in PNA content as they mature from green to yellow (Table XXXI). It is interesting that all PNAs increase by about the same factor in tree leaves (5-fold) and in tobacco (3-fold) with aging.

Biosynthesis of PNAs has also been investigated with respect to bacterial synthesis (Brisou 1969). Very small (trace) quantities of B(a)P were observed with a variety of substrates. The rate of appearance of 3,4-benzpyrene in marine phytoplankton was investigated by Niaussat *et al.* (1969). Presence of B(a)P was demonstrated as due to the bacteria.

However, careful investigations with higher plants indicate that the biosynthesis scheme is still questionable. Grimmer and Duvel (1970) demonstrated that absence (0 μg/kg) of benz(e)pyrene, perylene, anthanthrene, benzo(ghi)perylene, dibenz(a,h)anthracene, and coronene in higher plants in growth chambers with filtered air. These PNAs were present in low concentrations in duplicate plants grown under greenhouse conditions. Results are summarized in Table XXXII. The soils and concentration of PNAs in soils used for those experiments were not discussed.

d) Phytotoxic effects of PNAs

The only phytotoxic effects reported for PNAs include reduced growth of wheat and barley (Wagner and Wagner-Hering 1971). This information is summarized in Table XXVIII. No effect was observed on rye, summer wheat, and green maize at 3,4-benzpyrene and 3,4-benzfluoranthene (B(b)F) concentrations up to 1,200 μg/kg. Increasing soil B(b)fluoranthene concentration to 6,254 μg/kg reduced stem growth in summer wheat, but did not significantly affect rye plants. Dry leaf mass was slightly reduced in the case of plants with B(b)F, and total dry yield was reduced by 11% (36.5 to 32.5 g). Phytotoxic

Table XXXII. *PNA levels in higher plants grown in different environments (μg/kg)* (Grimmer and Duvel 1970).

PNA	Climate-controlled (results for all plants)	Greenhouse			
		Lettuce	Tobacco	Rye	Soybean
Benz(e)pyrene	0	4.2	2.5	3.4	4.3
Benz(a)pyrene	0	3.7	1.8	1.6	3.1
Perylene	0	0.4	0.25	0.9	0.3
Anthanthrene	0	0.2	0	0.1	trace
Benzo(ghi)perylene	0	2.5	1.1	0.9	1.5
Dibenz(a,h)anthracene	0	0.4	trace	0.1	0.1
Coronene	0 (0.1[a])	0.6	0.1	0.1	0.4

[a] 0.1 μg/kg for soybean only.

Table XXXIII. *Distribution of radioactivity among products of metabolism of ^{14}C-benzene in the tea plant* (Durmishidze and Ugrekhelidze 1969).

Part of plant	Organic acids								^{14}C as % of total radioactivitv of the fraction	amino acids							
	fumaric	succinic	malonic	malic	citric	glycollic	lactic	mandelic	dry residue[a]	leucine	valine	methionine	threonine	alanine	tyrosine	glycine	glutamic
	Assimilation of benzene through the roots																
Young roots		15.6	50	11.1					21.7	13	32	11.5	17.5			26	
Year-old roots	20	20	27	33						21.7	43.3	26					30
3-year roots	23	24.5	26.2	26.3						17.5	38.5		14				
Stem	88	12								18	24	20	14			2	22
Year-old shoots	64		10.5	25.5						17.1	22.2		23			14.2	8.5
	Assimilation of benzene through the stem																
Leaves	12.5		35		52.5					11.7	8	24		18.3	38		
Year-old shoots	30	70								7	28	21	7	3.5	33.5		
Stem above injection point	27	28.7	7.8			1.5	1.3	33.7		7	9.3	11.7	4.3	11.7	56		
Stem at injection point	7.1	51.4	21.4				3.4	16.7			9.4	28			62.6		
3-year roots	44	47.2	8.8							35	12.5	52.5					
Young roots	7.7	28.7	60								50	42.5	7.5				

[a] Dry residue from extraction with 80% ethanol.

effects are apparently not severe in these higher plants (wheat and barley) up to B(a)P and B(b)F soil concentrations of 18,000 μg/kg, even though it appears that B(b)F is bioaccumulated with the amount of uptake related to soil B(b)F concentration.

e) Plant metabolism of PNAs

Plants appear to be able to metabolize the benzene ring. Durmishidze and Ugrekhelidze (1969) demonstrated the cleavage of the carbon atom into organic acids and amino acids. Radioactive carbon was translocated from roots upward and from the stem downward and upward through the plant.

The first stable products of benzene metabolism by tea plants appear to be organic acids. These acids accumulate in the plant part where benzene was introduced (roots and stems).

A radioactivity balance indicated that radioactivity excreted as $^{14}CO_2$ (plants were kept in dark) during the experiment of 72 hr amounted to 3% for root application of benzene and 2% for stem application.

A proposed pathway for benzene degradation in plants was postulated:

benzene → phenol → pyrocatechol → *o*-benzoquinone → muconic acid

Table XXXIII presents the results for organic acid and amino acid distribution for benzene assimilated through roots and through the stem.

Durmishidze *et al.* (1973) studied the assimilation and conversion of labeled 3,4-benzpyrene-1,2-^{14}C with sterile corn and bean plants. The experiments were conducted on 14-day-old corn seedlings (*Zea mays*) and beans (*Phaseolus vulgaris*) cultivated under sterile conditions in a known nutrient medium. 3,4-Benzpyrene was introduced either through the roots or though the leaves. Results of the experiment, indicating the distribution of radioactivity in plant metabolism products are presented in Tables XXXIV and XXXV. Low molecular weight substances include organic acids, amino acids, and sugars. The dry residue not soluble in 80% ethanol represents the complex biopolymers.

The results show that higher plants are capable of metabolizing B(a)P when it is introduced to the roots or to the leaves. The quantity of unmetabolized B(a)P appeared to be a function of the length of time of the experiment. The largest fraction of radio-labelled carbon is associated with the low molecular weight substances, primarily organic acids. These are likely to be the primary products of biological degradation of B(a)P.

Durmishidze *et al.* (1974) extended their research on B(a)P metabolism to several additional plants including alfalfa (*Medicago sativa*), rye grass (*Lolium multiflorum*), chick-pea (*Cicer orientinum*), cucumbers (*Cucumis sativus*), pumpkin (*Cucurbita*), orchard grass (*Dactylis glome rata*), and vetch (*Vicia faba*). Results were similar to those obtained for corn and bean plants. Organic acids were most radioactive and were concentrated at the site of assimilation. For assimilation by roots, organic acid radioactivity ranged from 5.4 to 56.5% of

Table XXXIV. *Assimimilation of 3,4-benzpyrene-1,2-^{14}C by sterile seedlings of corn and beans (assimilation from roots, exposure 92 hr; temperature 18° to 22°C, in darkness)* (Durmishidze *et al.* 1973).

Plant	Plant organ	Radioactivity, thousand dpm/min on 1 g				
		Fraction of biopolymers	Fraction of low molecular wt substances + BP[a]	Fraction of organic acids	Fraction of amino acids	$^{14}CO_2$
Corn	roots	0.137	144.4	95.1	6	0.1
	leaves	0.364	8.95	4.2	1.15	–
Beans	roots	0.2	200.2	131.3	8.7	0.03
	leaves	0.11	7.1	3.25	1.1	–

[a] BP = 3,4-benzpyrene-1,2-^{14}C.

Table XXXV. *Assimilation of 3,4-benzpyrene-1,2-^{14}C by sterile seedlings of beans (assimilation from leaves, exposure 224 hr, temperature 19° to 22°C in darkness)* (Durmishidze *et al.* 1973).

Plant	Radioactivity, thousand dpm/min on 1 g				
	Fraction of biopolymers	Fraction of low molecular wt substances	Fraction of organic acids	Fraction of amino acids	$^{14}CO_2$
Beans (leaves and roots)	0.13	28.7	18.1	1.9	0.2

the total radioactivity of the root biomass. For assimilation by leaves the range was from 2.1 to 62.2% of the total radioactivity of the leaf biomass. Radioactivity was also incorporated into amino acids (up to 18% of total radioactivity) and $^{14}CO_2$ (up to 9%). The amount of 1,2-^{14}C-B(a)P metabolized ranged from 2 to 18% of the B(a)P assimilated by the plants.

VII. Conclusions

Polynuclear aromatic (PNA) compounds are a class of organics including hydrocarbon and heterocyclic species. Toxicity, metabolic and environmental fate, and environmental control of PNAs are topics of intense current study. Individual PNA compounds and isomers vary widely in their susceptibility to biodegradation; however, a consistent trend of decreasing biodegradation with increasing ring number is apparent. As shown, PNAs do decompose in soil systems, probably more rapidly than in any other waste management alternative or receiver systems involving biological activity. Research is needed to establish conclusively rates of decomposition and soil levels above which application of PNA constituents adversely affects the food chain. Thus utilization of a soil assimilative capacity concept will allow treatment of PNA compounds in an environmentally acceptable and cost effective manner. Research must be initiated to study enhancement factors that increase degradation rates of PNAs, especially those with more than three rings. Such research would not only add to present knowledge of PNA decomposition pathways by soil microorganisms under environmental conditions, but may also substantially increase the soil assimilative capacity and provide land application as an alternative technology that may be the most economical for safe and acceptable disposal of PNA wastes.

Based on the information presented, it appears that with bacterial seeding or in situ acclimation of a land application site, a substantial degradation rate could be achieved for the PNA class of compounds.

Of all degradation pathways considered, including volatilization and microbial degradation, the latter mechanism appears to be the most important in the environment.

Furthermore, natural regulation systems in the soil-plant cycle appear to keep plant uptake and biomagnification quite low in comparison to alternative stream receiver systems.

Storage and land-filling of PNA wastes are not ultimate solutions but simply postpone the problem of disposal; however, PNAs may be diluted in soil below threshold levels that are nontoxic and not a hazard to either food chain, water supplies, or agricultural productivity.

References

Ahokas, J., R. Paakkonen, K. Ronnholm, V. Raunio, and O. Pelkonen: Oxidative metabolism of carcinogens by trout liver resulting in binding and mutagenicity. Hoppe-Seyler's Z. Physiol. Chem. Bd. **357**, S: 1019 (1976).

Akhtar, M. N., D. R. Boyd, N. J. Thompson, D. T. Gibson, V. Mahadevan, and D. M. Jerina: Absolute stereochemistry of the dihydroanthracene-cis- and trans-1,2-diols produced from anthracene by mammals and bacteria. J. Chem. Soc. **1506** (1975).

Alexander, M.: Introduction to soil microbiology, 2nd ed. New York: Wiley (1977).

Ames, B. N., P. Sims, and P. L. Grover: Epoxides of polycyclic hydrocarbons are frameshift mutagens. Science **176**, 47 (1972)

Andleman, J. B., and M. J. Suess: The photodecomposition of 3,4-benz-pyrene sorbed on calcium carbonate. In Foust S. J., and J. U. Hunter (eds.): Organic compounds in aquatic environments. New York: Marcel Dekker, pp. 439-468 (1971).

Arffman, E., and B. C. Christensen: Studies on the newt test for carcinogenicity. I. Benzo(a)pyrene, dibenz(a,h)anthracene, and 3-methylcholanthrene. Acta Pathol. Microbiol. Scand. **52**, 330 (1961).

Bailey, E. J., and N. Dungal: Polycyclic hydrocarbons in icelandic smoked food. Brit. J. Cancer. **12**, 348 (1958).

Bailey, G. W., and J. L. White: Review of adsorption and desorption of organic pesticides by soil colloids, with implications concerning pesticide bioactivity. J. Agr. Food Chem. **12**, 324 (1964).

Balls, M.: Benzpyrene-induced tumors in the clawed toad, *Xenopus laevis*. Experientia **20**(3), 143 (1964).

Bartle, K. D., M. L. Lee, and M. Novotny: An integrated approach to the analysis of air-pollutant polynuclear aromatic hydrocarbons. Proc. Anal. Div. Amer. Chem. Soc., Oct., pp. 304-307 (1976).

Bazin, M. J., P. T. Saunders, and J. I. Prosser: Models of microbial interactions in the soil. CRC Critical Reviews in Microbiology, pp. 463-498 (1976).

Bigger, C. A. H., J. E. Tomaszewski, A. W. Andrews, and A. Dipple: Evaluation of metabolic activation of 7,12-dimethylbenz(a)anthracene in vitro by Araclor 1254-induced rat liver S-9 fraction. Cancer Res. **40**, 655 (1980).

Binet, L., and L. Mallet: Diffusion of pH in the animated environment. Gaz. Hop. (Paris) **135**, 1142 (1963).

Blum, S. C., and R. E. Swarbrick: Hydroponic growth of crops in solutions saturated with ^{14}C-benzo(a)pyrene. J. Agr. Food Chem. **25**, 1093 (1977).

Blumer, M.: Curtisite, idrialite, and pendletonite, polycyclic aromatic hydrocarbon minerals: Their composition and origin. Chem. Geol. **16**, 245 (1975).

______ Polycyclic aromatic compounds in nature. Sci. American **234**(3), 35 (1976).

______, and W. W. Youngblood: Polycyclic aromatic hydrocarbons in soils and recent sediments. Science **188**, 53 (1975).

Borneff, J., and R. Knerr: Carcinogenic substances in water and soil. II. Stability of 3,4-benzpyrene in light. Arch. Hyg. **143**, 405 (1959).

_______, and H. Kunte: Carcinogenic substances in water and soil. XIV. Further investigations concerning polycyclic aromatic hydrocarbons in soil samples. Arch. Hyg. **147**, 401 (1963).

_______, G. Farkazdi, H. Glathe, and H. Kunte: The behavior of polycyclic aromate hydrocarbons in experiments with garbage-sewage sludge composts as fertilizers. Zlb. Bakt. Hyg. **I**. Abt. Orig. B, 157:151 (1973).

_______, F. Selenka, H. Kunte, and A. Maximos: Experimental studies on the formation of polycyclic aromatic hydrocarbons in plants. Env. Res. **2**, 22 (1968 a).

_______ _______ _______ _______ The synthesis of 3,4-benzopyrene and other polycyclic aromatic hydrocarbons in plants. Arch. Hyg. **152**, 279 (1968 b).

Bourcart, J., and L. Mallet: Marine pollution of the shores of the central region of the Tyrrhenian Sea (Bay of Naples) by BP-type, pH. C. R. Acad. Sci. (Paris) **260**, 3729 (1965).

Boyland, E.: Studies in tissue metabolism. II. The inhibition of lactic dehydrogenase by derivatives of carcinogenic compounds. Biochem. J. **27**, 791 (1933).

Braunstein, H. M.: Environmental, health, and control aspects of coal conversion: An information overview. Oak Ridge National Laboratory. Prepared for Energy Research and Development Administration. ORNL/E15-94 (1977).

_______, E. D. Copenhaver, and H. A. Pfruderer (eds.): Environmental, health, and control aspects of coal conversion: an information overview. ORNL/EIS-94 (1977).

Brisou, J.: Biosynthesis of 3,4-benzpyrene and anaerobiosis. C. R. Soc. Biol. **163**, 772 (1969).

Brown, E. R., J. J. Hazdra, L. Keith, F. Greenspan, J. B. Kwapinski, and P. Beamer: Frequency of fish tumors found in a polluted watershed as compared to nonpolluted Canadian waters. Cancer Res. **33**, 189 (1973).

Calder, J. A., and J. H. Lader: Effect of dissolved aromatic hydrocarbons on the growth of marine bacteria in batch culture. Applied Environ. Microbiol. **32**, 95 (1976).

Cerneglia, C. E., and D. T. Gibson: Metabolism of naphthalene by cell extracts of *Cunninghamella elegans*. Arch. Biochem. Biophys. **186**, 121 (1978).

_______ _______ Oxidation of benzo(a)pyrene by the filamentous fungus *Cunninghamella elegans*. J. Biol. Chem. **254**, 12174 (1979).

_______, R. L. Herbert, R. H. Dodge, P. J. Szaniszlo, and D. T. Gibson: Some approaches to studies on the degradation of aromatic hydrocarbons by fungi. In A. L. Bourquin and H. Pritshard (eds.); Microbial degradation of pollutants in marine environments. EPA-600/9-79-012:360-369 (1979).

Colucci, J. M., and C. R. Begeman: Carcinogenic air pollutants in relation to automotive traffic in New York. Environ. Sci. Tech. **5**, 145 (1971).

Dagley, S.: Microbial degradation of organic compounds in the biosphere. Amer. Scientist **63**, 681 (1975).

_______, W. C. Evans, and D. W. Ribbons: Polycyclic aromatic compounds. Nature **188**, 560 (1960).

Dao, T. L., C. King, and T. Tominaga: Isolation, identification, and biological study of compounds derived from 3-methylcholanthrene by irradiation in dimethyl sulfoxide. Cancer Res. **31**, 1492 (1971).

Davies, I. W., R. M. Harrison, R. Perry, D. Ratnayake, and R. A. Wellings: Municipal incinerator as source of polynuclear aromatic hydrocarbons in environment. Environ. Sci. Tech. **10**, 451 (1976).

Dean-Raymond, D., and R. Bartha: Biodegradation of some polynuclear aromatic petroleum components of marine bacteria. Chapter 9, Vol. 16 of Development in industrial microbiology. Society for Industrial Microbiology. Amer. Inst. Biol. Sciences, Washington, D.C., pp. 97-109 (1975).

Diehl, E. K., F. du Brevil, and R. A. Glenn: Polynuclear emission from coal-fired installations. J. Eng. Power. **89**, 276 (1967).

Dobbs, R. A., and J. M. Cohen: Carbon adsorption isotherms for toxic organics, U.S. Environmental Protection Agency 600/8-80-023, 321 p. (1980).

Durmishidze, S. V., T. V. Devadariani, L. K. Kavtaradze, and L. S. H. Krartskhava: Assimilation and conversion of 3,4-benzypyrene by plants under sterile conditions. Doklady Academii Nauk SSSR **218**, 1468 (1974).

________, and D. S. H. Ugrekhlidze: Degradation of benzene in tea plants. Doklady Academii Nauk SSSR **184**, 228 (1969).

________ ________, T. V. Devadariani, and L. K. Kavtaradze: Cleavage of 3,4-benzpyrene-1,2-^{14}C by higher plants. Bull. Acad. Sci. Georgian SSR **70**, 460 (1973).

Eglinton, G. T., and M. T. Murphy: Organic geochemistry. New York: Pergamon (1968).

Ershova, K. P.: Studies of the content of polycyclic hydrocarbons in effluents of petro-chemical industry and surface waters. Hyg. Sanit. **36**, 474 (1971).

Erskine, R. L., and E. V. Whitehead: Composition and analysis of marine pollutants. Iranian J. Sci. Tech. **3**, 221 (1975).

Flaig, W.: Chemistry of humic substances in relation to coalification. In R. F. Gould (ed.): Coal science, Advances In Chemistry Series **55**, 58 (1965).

Gardner, W. S., R. F. Lee, K. R. Tenore, and L. W. Smith: Degradation of selected polycyclic aromatic hydrocarbons in coastal sediments: Importance of microbes and polychaete worms. Water, Air, and Soil Pollution **II**, 339 (1979).

Geissman, T. A., K. Y. Sim, and J. Murdoch: Organic minerals, picene, and chrysene as constituents of the mineral curtisite (idrialite). Experientia **23**, 793 (1967).

Gibson, D. T.: Microbial degradation of aromatic compounds. Science **161**, 1093 (1968).

______ The microbial oxidation of aromatic hydrocarbons. CRC Critical Rev. Microbiol. **1**, 199 (1971).

______ Microbial degradation of carcinogenic hydrocarbons and related compounds. Proc. Symp. Sources, Effects, and Sinks of Hydrocarbons in Aquatic Environment. Aug. 9-11, pp. 225-237 (1976).

______, D. M. Jerina, H. Yagi, and H. J. C. Yeh: Oxidation of the carcinogens benzo(a)pyrene and benzo(a)anthracene to dihydrodiols by a bacterium. Science **189**, 295 (1975).

______, G. E. Cardini, F. C. Maseles, and R. E. Kallio: Incorporation of oxygen-18 into benzene by *Pseudomonas putida*. Biochem. **9**, 1631 (1970).

______, J. R. Koch, and R. E. Kallio: Oxidative degradation of aromatic hydrocarbons by microorganisms. I. Enzymatic formation of catechol from benzene. Biochem. **7**, 2653 (1968).

Giger, W., and M. Blumer: Polycyclic aromatic hydrocarbons in the environment: Isolation and characterization by chromatography, visible, ultraviolet, and mass spectrometry. Anal. Chem. **46**, 1663 (1974).

Graf, W.: On the natural occurrence and importance of carcinogenic polycyclic aromatic hydrocarbons. Mediainische Klinik **60**, 561 (1965).

______, and H. Diehl: Concerning the naturally caused normal level of carcinogenic polycyclic aromatics and its cause. Arch. Hyg. **150**, 49 (1966).

______, and A. G. Haller: The phototoxic activity of carcinogenic polycyclic hydrocarbons and their degradation products in the biological test. Zentralbl. Bakteriol. Parasitenko. Infektionskr. Hyg. Abt. 1:Orig., Reihe B 164, 250 (1977).

______, and W. Nowak: Growth stimulation in lower and higher plants by carcinogenic polycyclic aromatic compounds. Arch. Hyg. Bakt. **150**, 513, ORNL/tr-4111 (1966).

______, and C. Winter: 3:4 Benzypyren in erdol. Arch. Hyg. **152**, 289 (1968).

Griest, W. H., and S. E. Herbes: Characterization of environmental distribution of polycyclic aromatic hydrocarbons in sediment and water in the vicinity of a coal coking plant. Amer. Chem. Soc. Meeting, Div. Environ. Chem., Anaheim, CA, Mar. 18-22 (1978).

Grigorenko, L. T., P. P. Dikun, I. A. Kalinina, A. N. Mironova, and V. P. Rzhekhin: 3,4-Benzpyrene content in sunflower and cotton seed oils. Prikl. Biokhim. Microbiol. **6**, 142 (1970).

Grimmer, G., and D. Duvel: Endogenous formation of polycyclic hydrocarbons in higher plants. 8. Carcinogenic hydrocarbons in the environment of humans. Aeitschrift Naturforsch Teil B. **256**, 1171 (1970).

______, and A. Hildebrand: Content of polycyclic hydrocarbons in various types of vegetables and lettuce. Deutsche Lebensmittel-Rundschau. Heft **8**, 237 (1965).

Groenewegen, D., and H. Stolp: Microbial breakdown of polycyclic aromatic hydrocarbons. Tbl. Bakt. Hyg. I. Abt: Orig. B162, 225 (1976).

Gubergrits, M., L. Paalme, and R. Kirso: Degradation of benzo(a)pyrene and phenol by physicochemical agents during self-purification of reservoirs, Vapr. Profil Zagryazeniya Vnesh Sreds, Chastnosti Vodoemor, Kentseragen Veshchestvami, pp. 49-53 (1972).

Hakama, M., and E. A. Saxen: Cereal consumption and gastric cancer. Int. J. Cancer **2**, 265 (1967).

Hancock, J. L., H. G. Applegate, and J. D. Dodd: Polynuclear aromatic hydrocarbons on leaves. Atmos. Environ. **4**, 363 (1970).

Hangebrauck, R. P., D. J. Von Lehmden, and J. E. Meeker: Emissions of polynuclear hydrocarbons and other pollutants from heat generation and incineration processes. J. Air. Pollut. Control Assoc. **14**, 267 (1964).

Harrison, R. M., R. Perry, and R. A. Wellings: Polynuclear aromatic hydrocarbons in raw, potable, and waste waters. Water Res. **9**, 331 (1975).

Hase, A., and R. A. Hites: On the origin of polycyclic aromatic hydrocarbons in recent sediments: Biosynthesis of anaerobic bacteria. Geochim. et. Cosmochim. Acta **40**, 1141 (1976).

Hass, B. S., and H. G. Applegate: The effects of unsubstituted polycyclic aromatic hydrocarbons on the growth of *Escherichia coli*. Chem. Biol. **10**, 265 (1975).

Hassett, J. J., J. C. Means, W. L. Banwart, S. G. Wood, A. Khan, and S. Ali: Sorption of dibenzothiophene by soils and sediments. J. Environ. Qual. **9**, 184 (1980).

Hattori, T., and R. Hattori: The physical environment in soil microbiology: an attempt to extend principles of microbiology to soil microorganisms. CRC Critical Rev. Microbiol., pp. 423-461 (1976).

Hayaishi, O., and M. Nozaki: Nature and mechanisms of oxygenases. Science **164**, 389 (1969).

Hayatsu, R., R. G. Scott, L. P. Moore, and M. H. Studier: Aromatic units in coal. Nature **257**, 378 (1975).

Herbes, S. E.: Partitioning of polycyclic aromatic hydrocarbons between dissolved and particulate phases in natural waters. Water Res. **11**, 493 (1977).

______ Rates of microbial transformations of polycyclic aromatic hydrocarbons in waters and sediments in the vicinity of a coal-coking wastewater discharge. Applied Environ. Microbiol. **41**, 20 (1981).

______, and L. R. Schwall: Microbial transformations of polycyclic aromatic hydrocarbons in pristine and petroleum-contaminated sediments. Applied Environ. Microbiol. **35**, 306 (1978).

______, G. R. Southworth, and C. W. Gehrs: Organic contaminants in aqueous coal conversion effluents: environmental consequences and research priorities. Trace substances in environmental health-X. A Symposium D. D. Hemphill (ed.), Univ. Missouri, Columbia (1976).

Hites, R. A., R. E. La Flamme, and J. W. Farrington: Sedimentary polycyclic aromatic hydrocarbons: The historical record. Science **19**, 829 (1977).

Hodgson, E., and F. E. Guthrie: Introduction to biochemical toxicology. New York: Elsevier (1980).

Hodgson, G. W., B. Hitshon, K. Taguchi, B. L. Baker, and E. Peake: Geochemistry of porphyrins, chlorins, and polycyclic aromatics in soils, sediments, and sedimentary rocks. Geochim. Cosmochim. Acta. **32**, 737 (1968).

Horton, A. W., M. J. Burton, R. Tye, and E. L. Bingham: Composition and contaminants of distillate oils. Amer. Chem. Soc. Div. Petrol. Chem. Preprint **8**, 59 (1963).

Hussien, Y. A., M. S. Tewfik, and Y. A. Hamdi: Degradation of certain aromatic compounds by rhizobia. Soil Bio. Biochem. **6**, 377 (1974).

Il'Nitskii, A. P.: Control of the pollution of water basins by carcinogenic hydrocarbons. Hyg. Sanit. **31**, 389 (1966).

______, and J. P. Cohan: Prophylaxis of food contamination with carcinogens. Voprosy Onhologii **18**, 106 (1972).

______, R. P. Ershova, A. Ya. Khesina, L. G. Rozhkova, V. G. Klubkov, and A. A. Korolev: Stability of carcinogenic substances in water and the efficacy of methods of decontamination. Gig. Sanit. **36**, 8 (1971).

______, L. G. Solenova, and V. V. Ignatova: Sanitary and oncological assessment of agricultural use of sewage containing carcinogenic hydrocarbons. Kazanskii Meditsinskii Zh. 2:80. ORNL/tr-2959 (1974).

———, V. Yu Gvil'dis, U. S. Muschenko, and L. M. Shabad: Role of volcanoes in the formation of the natural level of carcinogens. Doklady Akademii Nauk SSSR **234**, 717 (1977).

Ishio, S., H. Nakagawa, and T. Tomiyama: Algal cancer and its causes. II. Separation of carcinogenic compounds from sea bottom mud polluted by wastes from the coal chemical industry. Bull. Japan Soc. Sci. Fish. **38**, 571, ORNL/tr-2848 (1972).

Jackson, H. M.: Synthetic fuels from coal: status and outlook of coal gasification and liquefaction. Committee on Energy and Natural Resources, U.S. Senate, Pub. No. 96-17, 1-196 (1979).

Jamison, V. W., R. L. Raymond, and J. O. Hudson: Microbial hydrocarbon cooxidation. III. Isolation and characterization of an α,α',dimethyl-cis, cismuconic acid-producing strain of *Norcardia corallina*. Applied Microbiol. **17**, 853 (1969).

Jeffrey, A. M., H. J. C. Yeh, D. M. Jerina, T. R. Patel, J. F. Davey, and D. T. Gibson: Initial reactions in the oxidation of naphthalene by *Pseudomonas putida*. Biochem. **14**, 575 (1975).

Jerina, D. M., J. W. Daly, A. M. Jeffrey, and D. T. Gibson: Cis-1,2-dihydroxy-1,2-dihydronaphthalene: a bacterial metabolite from naphthalene. Arch. Biochem. Biophys. **142**, 394 (1971).

Jones, J. R. E.: Fish and river pollution. London: Butterworths (1964).

Junk, G. A., and C. S. Ford: A review of organic emissions from selected combustion processes. Chemosphere **9**, 187 (1980).

Karickhoff, S. W., D. S. Brown, and T. A. Scott: Sorption of hydrophobic pollutants of natural sediments. Water Res. **13**, 241 (1979).

Karu, T., U. Kirse, and M. Gubergrits: Kinetics of the photoinduced cooxidation of benzo(a)pyrene and phenols. Eesti NSV Tead Akad. Toim, Keem, Geol. **22**, 217 (1973).

Khesina, A. Ya, N. P. Shcherback, L. M. Shabad, and I. S. Vostrov: Benzpyrene breakdown by the soil microflora. Byulleten Eksperimental'noi Biologii i Meditsiny **68**, 70 (1969).

Kingsbury, G. L., R. C. Sims, and J. B. White: Multimedia goals for environmental assessment. EPA-600/7-79-176b (1979).

Kolar, L., R. Ledvina, J. Ticha, and R. Hanus: Contamination of soil, agricultural crops, and vegetables by 3,4-benzpyrene in the vicinity of Ceska Budejovice. Ceskosolvenska Hygiena **20**, 135 (1975).

Lane, D. A., and M. Katz: The photomodification of benzo(a)pyrene, benzo(b)-fluoranthene, and benzo(k)fluoranthene under simulated atmospheric conditions. In Suffet I. H. (ed.): Fate of pollutants in the air and water environments. Part 2, pp. 135-154. New York: Wiley (1977).

Lee, R., W. S. Gardner, J. W. Anderson, J. W. Blaylock, and J. B. Clarke: Fate of polycyclic aromatic hydrocarbons in controlled ecosystem enclosures. Environ. Sci. Tech. **12**, 832 (1978).

Lijinsky, W., and J. H. Quastel: Metabolism of carcinogenic hydrocarbons by soil microorganisms. Arch. Biochem. Biophys. **63**, 160 (1956).

Lopp, A., L. Paalme, and M. Gubergrits: Kinetics of photoinduced oxidation of 7,12-dimethylbenz(a)anthracene in various solvents. Eesti NSV Tead Akad. Toim, Keem, Geol. **25**, 22 (1976).

Lu, P., R. L. Metcalf, N. Plummer, and D. Mandel: The environmental fate of three carcinogens: Benz(a)pyrene, benzidine, and vinyl chloride evaluated in laboratory model ecosystems. Arch. Environ. Contam. Toxicol. **6**, 129 (1977).

Mackay, D., and P. J. Leinonen: Rate of evaporation of low-solubility contaminants from water bodies to atmosphere. Environ. Sci. Tech. **9**, 1178 (1975).

______, and A. W. Wolkoff: Rate of evaporation of low-solubility contaminants from water bodies to atmosphere. Environ. Sci. Tech. **7**, 611 (1973).

Macphee, C., and R. Ruelle: Lethal effects of 1888 chemicals upon four species of fish from western North America. Univ. Idaho, Coll. Forestry, Wildlife and Range Sciences, Bull. No. 3 (1969).

Mailath, F. P., F. Medve, and J. Morik: Photolysis of 3,4-benzypyrene by ultraviolet radiation and sunlight. Egeszsgtudomany **13**, 333 (1974).

Mair, B. J., and J. L. Martinez-Pico: Diagenesis of plant sterols. Proc. Amer. Petrol. Inst. **42**, 173 (1962).

Malaney, G. W., P. A. Lutin, J. J. Chibulka, and L. H. Hickerson: Resistance of carcinogenic organic compounds to oxidation by activated sludge. J. Water Pollut. Control Fed. **39**, 20202 (1967).

Mallet, L., A. Perdriau, and J. Perdriau: The extent of pollution by BP-type PH in the North Sea and Glacial Arctic Ocean. Bull. Acad. Nat. Med. (Paris) **147**, 320 (1963).

______, J. Zanghi, and J. Brisou: Investigations of the possibilities of biosynthesis of a polybenzene hydrocarbons of the benzo-3,4-pyrene type by a *Clostridium putride* in the presence of marine plankton lipids. Acad. Sci., Paris Ser. D. **264**, 1534 (1967).

McBride, M. B., T. J. Pinnavaia, and M. M. Mortland: Adsorption of aromatic molecules by clays in aqueous suspensions. Ithaca, New York: N.Y. State College of Agriculture and Life Sci., Cornell University (1975).

McCann, J., and B. N. Ames: A simple method for detecting environmental carcinogens as mutagens. Ann. N.Y. Acad. Sci. **271**, 5 (1975).

McGinnes, P. R., and V. L. Snoeynik: Determination of the fate of polynuclear aromatic hydrocarbons in natural water systems. WRC Res. Rep. 80. Urbana, Illinois: University of Illinois Water Resources Center (1974).

McKee, J. E., and H. W. Wolf (eds.): Water quality criteria, 2nd ed., Calif. State Water Resources Control Board (1963).

McKenna, E. J.: Biodegradation of polynuclear aromatic hydrocarbon pollutants: By soil and water microorganisms. Amer. Inst. Chem. Eng. 70th Ann. Meet. Nov. 13-17, New York, NY (1977).

Means, J. C., J. J. Hassett, S. G. Wood, and W. L. Banwart: Sorption properties of energy-related pollutants and sediments. 327-340. In P. W. Jones and P. Leber (eds.): Polynuclear aromatic hydrocarbons. Ann Arbor, Mich.: Ann Arbor Sci. Publ., Inc. (1979).

______, G. S. Wood, J. J. Hassett, and W. L. Banwart: Sorption of polynuclear aromatic hydrocarbons by sediments and soils. Environ. Sci. Tech. **14**, 1524 (1980).

Medvedev, V. A., and V. D. Davidov: Transformation of individual organic products of the coke industry in chernozemic soils. Pochvovedenie **11**, 22 (1972).

Murdock, J., and T. A. Geissman: Pendletonite, a new hydrocarbon mineral from California. Amer. Mineral **52**, 611 (1967).

National Academy of Sciences, National Research Council: Particulate polycyclic organic matter. Washington, D.C.: National Academy of Sciences (1972).

Niaussat, P., C. Auger, and L. Mallet: Appearance of carcinogenic hydrocarbons in pure *Bacillus hadius* cultures relative to the presence of certain compounds in the medium. Compt. Rend. **270D**, 1042 (1970).

______, L. Mallet, and J. Ottenwaelder: Appearance of 3,4-benzpyrene in various strains of marine phytoplankton raised *in vitro*. Compt. Rend. Acad. Sci. Paris. t. **268**, 1109 (1969).

Nozaki, M., H. Kagamiyama, and O. Hayaishi: Metapyrochatechase. I. Purification, crystallization and some properties. Biochem. Z. **338**, 582 (1963).

Paalme, L., and M. Gubergrits: Kinetics of photodecomposition of 1,2-benzpyrene. Eesti NSV Tead Akad. Toim, Keem, Geol. **21**, 48 (1972).

______ ______ Kinetics of cooxidation of 1,2- and 3,4-benzpyrene activated by UV radiation. In N. M. Emanuel (ed.): Tear Prakt Zhidkofazn Okieleniya, pp. 12-122, Nauka, Moscow, U.S.S.R. (1974).

______ ______ Kinetics of the separate and combined photodegradation of benzo(a)pyrene, pyrene, and 3-methylcholanthrene. Eesti NSV Tead Akad. Toim, Keem, Geol. **25**, 271 (1976).

______, A. Lopp, and M. Gubergrits: Kinetics of cooxidative photodegradation of benzo(a)pyrene and 7,12-dimethylbenz(a)anthracene. Eesti NGV Tead Akadd. Toim, Keem, Geol. **25**, 247 (1976).

______, R. Priiman, M. I. Gloshko, and M. Gubergrits: Simultaneous photo-induced oxidation of 3,4-benzpyrene and alkyl arylsulfonates. Vodn. Resur., pp. 174-178 (1975).

Panalaks, T.: Determination and identification of polycyclic aromatic hydrocarbons in smoked and charcoal-broiled food products by high pressure liquid chromatography. J. Environ. Sci. Health **B 11**, 299 (1976).

Perry, J. J.: Microbial cooxidations involving hydrocarbons. Microbiol. Rev. **45**, 59 (1979).

Poglazova, M. N., G. E. Fedoseeva, A. J. Khesina, M. N. Meissel, and L. M. Shabad: Destruction of benzo(a)pyrene by soil bacteria. Life Sciences **6**, 1053 (1967 a).

______ ______ ______ ______ ______ Further investigations of the decomposition of benz(a)pyrene by soil bacteria. Doklady Akademii Nauk SSSR. **176**, 1165 (1967 b).

______ ______ ______ ______ ______ The oxidation of benz(a)pyrene by microorganisms in relation to its concentration in the medium. Doklady Akademii Nauk SSSR **179**, 1460 (1968).

Radding, S. B., T. Mill, C. W. Gould, D. H. Liu, H. L. Johnson, D. C. Bomberger, and C. V. Fojo: The environmental fate of selected polynuclear aromatic compounds. U.S. Environmental Protection Agency, 122 p. EPA-560/5-75-009 (1976).

Raymond, R. L., V. W. Jamison, and J. O. Hudson: Microbial hydrocarbon co-oxidation. I. Oxidation of mono- and dicyclic hydrocarbons by soil isolates. Applied Microbiol. **15**, 857 (1967).

Reinbold, K. A., J. J. Hassett, J. C. Means, and W. L. Banwart: Adsorption of energy-related organic pollutants: a literature review. Inst. Environ. Sci., Dept. Agron., Univ. Ill., Urbana, Champaign, for U.S. EPA Office for Research and Development, Athens, GA, 130 p. (1978).

Rogoff, M. H., and I. Wender: The microbiology of coal. I. Bacterial oxidation of phenanthrene. J. Bacteriol. **73**, 264 (1957).

Rohrlich, M., and P. Suckow: Oxidative modification of 3,4-benzpyrene on cereals and in pastry. Chem. Mikrobiol. Technol. Lebensm. **2**, 137 (1973).

Rondia, D., and S. S. Epstein: Effect of antioxidants on photodecomposition of benzo(a)pyrene. Life Sci. **7**, 513 (1968).

Ryan, P. W., and C. K. McMahon: Some chemical and physical characteristics of emissions from forest firest. Paper 76-2.3, 69th Annual Meeting Air Pollution Control Assoc., Portland, OR, June 27-July 1 (1976).

Sawicki, E., J. E. Meeker, and M. J. Morgan: Polynuclear aza compounds in automotive exhaust. Arch. Environ. Health **11**, 773 (1965).

Selander, H., H. Yagi, D. M. Jerina, M. C. Wells, J. F. Davey, V. Mahadevan, and D. T. Gibson: Dihydrodiols from anthracene and phenanthrene. J. Amer. Chem. Soc. **129**, 67 (1971).

Shabad, L. M.: Studies in the U.S.S.R. on the distribution, circulation, and fate of carcinogenic hydrocarbons in the human environment and the role of their disposition in tissues in carcinogenesis. A Review. Cancer Res. **27**, 1132 (1967).

______ Circulation of carcinogenic substances in the environment: From laboratory experiments to field investigations. GANN Monograph on Cancer Research **17**, 179 (1975).

______, and Y. L. Cohan: The contents of benzo(a)pyrene in some crops. Arch. Geschwultsforsch **40**, 237 (1972).

______ ______, A. Ya Khesina, H. P. Schubak, and G. A. Smirnov: The carcinogenic hydrocarbon benzo(a)pyrene in the soil. J. Nat. Cancer Inst. **47**, 1179 (1969).

Shamsuzzaman, K. M., and E. A. Barnsley: The regulation of naphthalene metabolism in pseudomonas. Biochem. Biophys. Res. **60**, 592 (1974).
1179 (1969).

Shamsuzzaman, K. M., and E. A. Barnsley: The regulation of naphthalene metabolism in pseudomonas. Biochm. Biophys. Res. **60**, 592 (1974).

Shcherbak, N. P.: Some results of a study of the fate of benz(a)pyrene in soil. Vapr. Onkel. **15**, 75 (1969).

Sherrill, T. W., and G. S. Sayler: Phenanthrene biodegradation in freshwater environments. Applied Environ. Microbiol. **39**, 172 (1980).

Shiraishi, Y.: Determination of polycyclic aromatic hydrocarbons in food. IV. 3,4-Benzpyrene in fish and shellfish. J. Food. Hyg. Soc. Japan **16**, 178 (1975).

______, and E. Takabatake: Determination of polycyclic aromatic hydrocarbons in foods. 3. 3,4-Benzopyrene in vegetables. Shokuhin Eiseigaku Zasshi **15**, 18, ORNL/tr-4125 (1974).

Siddiqi, I., and K. H. Wagner: Determination of 3,4-benzpyrene and 3,4-benzfluoranthene in rain water, ground water, and wheat. Chemosphere **1**, 83 (1972).

Siegfried, R.: Effect of garbage compost on the 3,4-benzpyrene content of carrots and head lettuce. Naturwissenschaften **62**, 300, ORNL/tr-4124 (1975).

Sims, R. C.: Land treatment of polynuclear aromatic compounds. PhD dissertation. Dept. Biol. Agr. Eng., No. Carolina State Univ., Raleigh, NC (1982).

Simmon, V., and J. M. Baden: Mutagenic activity of vinyl compounds and derived epoxides. Mut. Res. **78**, 227 (1980).

Sisler, F. D., and C. E. Zobell: Microbial utilization of carcinogenic hydrocarbons. Science **106**, 521 (1947).

Skrigan, A. I.: Diagenesis of α-pinene in pine stumps. Doklady Akad. Nauk. SSSR **80**, 607 (1951).

______ Diagenesis of abietic acid in buried pine trees. Through Chem. Abstr. **62**, 10664 (1965).

Smirnov, G. A.: The study of the benz(alpha)pyrene content in the soil and vegetation in the airfield region. Vop. Onkol. **16**, 83 (1970).

Smith, J., W. R. Mabey, N. Bohomos, B. R. Holt, S. S. Lee, T. W. Chou, D. C. Bomberger, and T. Hill: Environmental pathways of selected chemicals in freshwater systems. U.S. EPA (Office of Research and Development), Athens, GA. EPA-600/7-78-074 (1978).

Southworth, G. R.: Transport and transformation of anthracene in natural waters: Process rate studies. Oak Ridge, Tennessee: U.S. Department of Energy (ORNL) 26 p. (1977).

______, J. J. Beauchamp, and P. K. Schmieder: Bioaccumulation potential of polycyclic aromatic hydrocarbons in *Daphnia pulex*. Water Res. **12**, 973 (1978).

Stanier, R. Y., and J. L. Ingraham: Protocatechuic acid oxidase. J. Biol. Chem. **210**, 799 (1954).

Stevens, B., and B. E. Algar: Photoperoxidation of unsaturated organic molecules. II. Autoperoxidation of aromatic hydrocarbons. J. Phys. Chem. **72**, 3468 (1968).

Stotzky, G.: Techniques to study interactions between microorganisms and clay minerals *in vivo* and *in vitro*. Bull. Ecol. Res. Comm. (Stockholm) **17**, 17 (1973).

Suess, M. J.: Aqueous solutions of 3,4-benzpyrene. Water Res. **6**, 981 (1972 a).

______ Polynuclear aromatic hydrocarbon pollution of the marine environment. In M. Ruivo (ed.): Marine pollution and sea life. London. Fishing New Books (1972 b).

______ The environmental load and cycle of polycyclic aromatic hydrocarbons. The Sci. of the Total Environ. 6:239-250 (1976).

Tausson, V. O.: In N. A. Meximov (ed.): Basic principles of plant bioenergetics. Acad. Sci., U.S.S.R. (1950).

Thakker, D. R., W. Levin, H. Yagi, M. Tada, A. H. Conney, and D. M. Jerina: Comparative metabolism of dihydrodiols of polycyclic aromatic hydrocarbons to bay region diol epoxides. Fourth Internat. Symp. Polynuclear Aromatic Hydrocarbons. Columbus, OH, Oct. 2-4, 1979, U.S. EPA: Battelle Press (1980).

Tye, R., M. J. Burton, E. Bingham, A. Bell, and A. W. Horton: Carcinogens in a cracked petroleum residuum. Arch. Environ. Health **13**, 202 (1966).

U.S. EPA: Review and evaluation of available techniques for determining persistence and routes of degradation of chemical substances in the environment. Syracuse Univ. Res. Corp., NTIS PB-243825, 549 pp. (1975 a).

______ Scientific and technical assessment report on particulate polycyclic organic matter (PPOM). Star Series. Superintendent of Documents, Washington, D.C.: U.S. Government Printing Office, EPA-600/6-74-001 (1975 b).

______ Rationale for recommended list of priority pollutants. Memo to director of Effluent Guidelines Division. Nov. 19 (1976).

______ Environmental pathways of selected chemicals in freshwater systems. Part II. Laboratory studies. EPA-600/7-78-074 (1978).

Versar, Inc.: Water related environmental fate of 129 priority pollutants. A literature search. V. polycyclic aromatic hydrocarbons, PCBs and related compounds. For U.S. EPA Office of Water Planning and Standards, Washington, D.C. (1979).

Wagner, Van K. H., and I. Siddiqi: The metabolism of 3,4-benzpyrene and 3,4-benzfluoranthene in summer wheat. Z. Pflanzen Bodenkundle **127**, 211 (1970).

Wagner, K. H., and C. Vonderheid: Polycyclic aromatic hydrocarbons in technically prepared composts. Naturwissenschaften **65**, 491 (1978).

______, and E. Wagner-Hering: The cycle of cancer-causing substances–polycyclic aromatic hydrocarbons–in plants and humans. Protectio Vitae **6**, 260 (1971).

Webley, D. M., R. B. Duff, and V. C. Farmer: The metabolism of iron-, zinc-, and manganese deficient bacteria. J. General Microbiol. **29**, 179 (1962).

Windsor, J. G., and R. A. Hites: Polycyclic aromatic hydrocarbons in Gulf of Maine sediments and Nova Scotia soils. Geochim et. Cosmochim. Acta **43**, 27 (1979).

Wiser, W. H.: Some chemical aspects of coal liquefication. Paper presented Short Course on Coal Characteristics and Coal Conversion Processes, Oct. 29-Nov. 2, Pennsylvania State Univ., Univ. Park, PA (1973).

Wright, A. S.: The role of metabolism in chemical mutagenesis and chemical carcinogenesis. Mut. Res. **75**, 215 (1980).

Wynder, E. L., J. Kmet, N. Dungal, and N. Segi: An epidemiological investigation of gastric cancer. Cancer **16**, 1461 (1963).

Youngblood, W. W., and M. Blumer: Polycyclic aromatic hydrocarbons in the environment: Homologous series in soils and recent marine sediments. Geochim. et Cosmochim. Acta **39**, 1303 (1975).

Manuscript received August 3, 1982; accepted October 24, 1982.

Photodynamic insecticides: A review of studies on photosensitizing dyes as insect control agents, their practical application, hazards, and residues

By
J. R. Robinson*

Contents

I. Introduction: Scope and purpose

The study of chemical compounds which exhibit light-sensitizing insecticidal activity is a complex and rapidly expanding field. There is increasing evidence that light-initiated toxic mechanisms play an important role in evolutionary processes as well as in natural control of insect populations. Graham (1963) was early in calling attention to the possibility of using "photosensitizing agents" as insecticides and later (1972 a) he enlarged on this in a discussion of the entomo-

*Agriculture Canada, Research Centre, University Sub Post Office, London, Ontario, Canada, N6A 5B7.

Residue Reviews, Volume 88.

logical, ecological and evolutionary implications of "photodynamic action." The latter is recognized today to be only one of the photochemical routes which are involved in biochemical and physiological damage caused by light; it is therefore necessary to limit the scope of this review.

The term photodynamic action (PDA) was narrowly defined (see below), one of the requisites being the presence of visible light, *i.e.*, electromagnetic radiation having wavelengths in the approximate range 390 to 740 nm. Photosensitizers which absorb a portion of this spectrum and transmit the remainder as a color are loosely termed "dyes" even though they may be useless as such in commerce. Thus PDA, by definition, is dependent upon a dye and visible light. Other processes, based upon the absorption of wavelengths outside the visible range by compounds which appear colorless are sufficiently significant, and mechanistically different from PDA, to provide the subject of a separate review; these involve a number of naturally-occurring sensitizers. Even among the events which are initiated in visible light by dye-sensitization there are some which do not involve the dynamic uptake of oxygen (*e.g.*, photochemical, direct free-radical mechanisms) and these, too, lie outside the scope of this work. Many of these mechanisms have been brought into modern perspective at a recent special symposium sponsored by the American Chemical Society (Maugh 1982).

Except for historical information necessary for definition of terms and for toxicological reports which appear relevant, this review is limited to studies of the effects of synthetic photosensitizing dyes on insects.

From 1928 to mid-1982 there have been 30 publications describing studies specifically designed to investigate the insecticidal activity of various dyes in the presence of visible light. The research objectives were to evaluate effectiveness, to elucidate the mechanism of PDA toxicity, and to explore the feasibility of applying this process in practical measures for insect control. The purpose of this review is to consolidate the bibliography relevant to the topic and to bring into focus the progress of the work on insects, most of which has occurred within the past decade. The summaries are intended to indicate only the highlights of results in each paper and the original reports should be consulted for details. The review provides, also, an opportunity for an independent assessment of possible problems with toxicology and residues which may be associated with future attempts to exploit the insecticidal photodynamic activity of dyes.

II. Photodynamic action (PDA)

a) Historical background and definition

Working with seeds, plants, fermentations, and amphibian eggs, Marccaci (1888) observed qualitatively that certain alkaloids had a more pronounced effect on protoplasm in daylight than in the dark. This appears to be the first formal report that some materials have a light-enhanced toxic effect, even though a few dyes were already being used for their anti-bacterial properties without recognition of the involvement of light. In 1900 Raab, after systematic quantitative studies of the effects of dyes (including acridine and eosin) on *Paramecium caudatum*, showed that the presence of daylight caused an increase in toxicity of 2 to 3 orders of magnitude in acridine solution at 50 ppm. He observed also that those light rays which generated the greatest "*Fluorescenz*" in the dyes also induced the largest biological reaction. Soon Jodlbauer and Tappeiner (1904, 1905) had proven that the presence not only of light but also of oxygen, and of active oxygen consumption in the process, were equally necessary to this toxic phenomenon in bacteria, enzymes, and toxins. They recognized that some activated form of oxygen was involved but could find no support for the then-current "peroxide theory." These authors (Tappeiner and Jodlbauer 1904) also introduced the new name "photodynamic action" to differentiate this activity from other dye-sensitized photo-phenomena such as inorganic photographic processes. Since that time the term has been restricted to dye-sensitized photo-oxidation of biological substrates with consumption of molecular oxygen. After seven decades in use, during which time this concept has been repeatedly challenged as new information became available, it has remained virtually unchanged (Table I); the only concession has been a broadening of the meaning of "biological substrate" to include not only intact organisms *in vivo* but also isolated "molecules of biological importance" such as the nucleic acids and their components, proteins, amino acids, enzymes (Spikes 1968, Spikes and Livingstone 1969), and cell membranes (Valenzeno and Pooler 1982, Barratt *et al.* 1982).

1. **Photosensitizing dyes.**–The dyes most frequently cited in this review possess molecular structures as illustrated in Figure 1. The dihydroxyxanthenes (fluorones) are in the closed (lactone) form (A) as free acids but rearrange to the open tautomeric structure (B) when in alkaline medium as anions, or as metal salts; fluorescein and all of its halogenated derivatives belong to this class. In the systematic nomenclature of *Chemical Abstracts* these dyes are all indexed as derivatives of the closed form, whether free acids or metal salts. The aminoxanthenes (fluorenes), having no oxy-functions at the 3- and 6-positions, are not capable of tautomerization and are therefore named as derivatives of entirely different parent structures; even within the class, the nomenclature of rhodamine 6G (an ester) is based upon a different parent molecule than that of rhodamine B [a free acid (C)]. Methylene blue (D) is a quinone-imine dye of the thiazine class and is named as a derivative of phenothiazine. For assistance in searching

Table I. *The definitive requisites of "photodynamic action."*

1. A biological substrate which suffers photooxidative damage in the process.
2. A source of visible light.
3. A photosensitizer ("dye") with maximum absorption in the range 400 to 700 nm and high phosphorescence emission.
4. Consumption of oxygen–usually molecular oxygen.

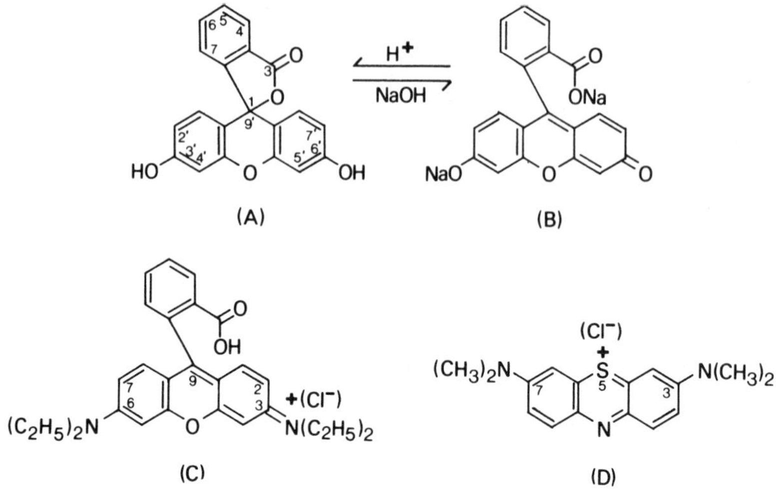

Fig. 1. Molecular structures of the fluoresceins, rhodamine B and methylene blue; refer to text and to Table II.

the literature, the Chemical Substance Index names are listed in Table II. Useful sources of more detailed information on the dyes, their chemical, physical, and optical properties, assay and purification methods, uses and suppliers, include Lillie (1969), Gurr (1971), and Marmion (1979).

b) Mechanism of PDA toxicity

The mechanism of PDA injury appears to result mainly, if not exclusively, from bio-oxidation at vital sites by the highly-activated singlet state of molecular oxygen $(^1O_2)$[1] (Nilsson *et al.* 1972, Kepka and Grossweiner 1973). The lower electronic states of excited oxygen are readily attained through the absorption of energy transferred from visible light via dye-sensitization. Of all dye types, the halogenated fluoresceins, a class within the "xanthene group" of dyes, were shown to be the most effective of oxygen sensitizers (Spikes and Glad 1964) and the relative efficiencies within the class have been measured using various substrates (Wade and Spikes 1971, Valenzeno and Pooler 1982). Callaham *et al.* (1975 a, 1977 b) have reviewed and diagrammatically illustrated the photoenergetic mechanism in the formation of $^1O_2$, showing the direct relationship

[1] Singlet oxygen has been characterized in a number of well-defined electronic states (Kasha and Brabham 1979). The longest-lived species ($O_2\,^1\Delta g$) is probably the most significant form in photodynamic activity as it has more time to enter into bioreactions; Merkel and Kearns (1972) list the mean lifetimes of singlet oxygen in a number of solvents. Throughout this review the term $^1O_2$ is used to denote "singlet molecular oxygen" in any of its effective forms.

Table II. *Names and index numbers of some photosensitizing dyes.*

Common name (synonyms)	Color index no.[a]	C.A.S. reg. no.[b]	C.A. Chemical Substance Index Name[c]	Mol. wt. (halogen content, %)[d]
Group: Xanthene				
Class: Fluorone				
1. Rose bengal (C.I. Acid Red 94, Rose bengale B, Tetraiodotetrachloro-fluorescein)	45440	632-69-9 (11121-48-5)	Spiro[isobenzofuran-1 (3H),9'-[9H] xanthen]-3-one, 4,5,6,7-tetrachloro-3',6'-dihydroxy-2',4',5',7'-tetraiodo, disodium salt	1,018. (63.8)
2. Erythrosine (erythrosin B, EBS, FD&C Red No. 3,[e] Ext. D&C Red No. 5, C.I. Food Red 14, C.I. Acid Red 51, Tetraiodofluorescein)	45430	16423-68-0	——— ——— 3',6'-dihydroxy-2',4',5',7'-tetra-iodo, disodium salt	880. (57.7)
3. Phloxine B (D&C Red No. 28, C.I. Acid Red 92, Tetrabromo-tetrachlorofluorescein)	45410	18472-87-2	——— ——— 2',4',5',7'-tetrabromo-4,5,6,7-tetrachloro-3',6'-dihydroxy, disodium salt	830. (55.6)

Table II. (*continued*)

	Common name (synonyms)	Color index no.[a]	C.A.S. reg. no.[b]	C.A. Chemical Substance Index Name[c]	Mol. wt. (halogen content, %)[d]
4.	Eosin Y (D&C Red No. 22, C.I. Acid Red 87, C.I. Pigment Red 90, Tetrabromofluorescein)	45380	17372-87-1	______ ______, 2′,4′,5′,7′-tetrabromo-3′,6′-dihydroxy, disodium salt	692. (46.2)
5.	Uranine (fluorescein, water-soluble, D&C Yellow No. 8 C.I. Acid Yellow 73)	45350	518-47-8	______ ______, 3′,6′-dihydroxy, disodium salt	376. (0)
6.	Fluorescein (Fluorescein acid, fluorescein insoluble, D&C Yellow No. 7 C.I. Solvent Yellow 94)	45350:1	2321-07-5	______ ______, 3′,6′-dihydroxy (free acid)	332. (0)
Group:	*Xanthene*				
Class:	*Fluorene*				
7.	Rhodamine B (C.I. Basic Violet 10, D&C Red No. 19, C.I. Food Red 15).	45170	81-88-9	Ethanaminium,*N*-[9-(2-carboxyphenyl)-6-(diethylamino)-3H-xanthen-3-ylidene]-*N*-ethyl, chloride (free acid)	479.
8.	Rhodamine 6G (C.I. Basic Red 1)	45160	989-38-8	Benzoic acid, 2-[6-(ethylamino)-3-(ethylimino)-2,7-dimethyl-3H-xanthen-9-yl]-,ethyl ester, monohydrochloride	479.

Group: Quinone-imine
Class: thiazin

9. Methylene blue (C.I. Basic blue, 9, Swiss blue, methylthionine chloride)	52015	61-73-4	Phenothiazin-5-ium,3,7-bis(dimethylamino)-,chloride	320

[a] Ref. Colour Index (1971).

[b] *Chemical Abstracts* Service Registry numbers. With the exception of fluorescein, all of the fluorones are listed as numbers assigned to the disodium salts; numbers differ for other salts and for parent acids although the same common name may be used.

[c] Refer to Figure 1 for numbering.

[d] Note that the fluorones, listed in descending order of molecular weight (and halogen content), are also in descending order of typical relative PDA-toxicities to insects (*e.g.*, to the face fly (Fondren and Heitz 1978 b)).

[e] F,D,C indicates usage in coloring foods, drugs or cosmetics (not necessarily currently permitted in every case).

between the mass of halogen-substitution in the fluorescein molecule, population of the triplet state of the dye sensitizer, and phosphorescence emission, with transfer of energy to oxygen resulting in its excitation to the singlet state.

The chemical and physical characteristics of $^1O_2$, its generation, electronic forms, and reactivity have been extensively reviewed. The monograph "Singlet Oxygen" (Wasserman and Murray 1979) contains several chapters with information relevant to its relationship to PDA which is not the only activity of this highly active species. Foote (1976) presents a precise delineation of the various photosensitized oxidation processes, including that of $^1O_2$ as opposed to free-radical systems, in several selected biological studies. In a wider-ranging review of photosensitization in biological substrates, Spikes (1977) describes all of those examined to that date with clear differentiation of the photochemical mechanisms involved. In addition to the extensive bibliographies cited above, more recent research- and review-studies include those of Phillips and Phillips (1981), Amagasa (1981), Rossi *et al.* (1981), Lochmann and Micheler (1979), Pooler and Valenzeno (1979), and the collection of conference papers edited by Singh and Petkau (1978).

Essentially all of these fundamental studies pointing to the implication of $^1O_2$ in PDA toxication have been done on substrates other than insects and they include plants, fungi, bacteria, viruses, protozoa, small mammals, fish, amphibia, birds, isolated cells, cell organelles, proteins, enzymes, nucleotides, amino acids, and others. In the 80 years since the defining of PDA there have been only seven publications (indicated in Table III) in which some insect biochemical parameters were measured in an attempt to clarify PDA mode-of-action; the results of these tend to confirm that singlet oxygen is the active toxic agent here just as in non-insect substrates.

Most evidence to date suggests that, rather than involving one specific critical organ as the site of toxic action, PDA damage *in vivo* may occur wherever an efficient photosensitizer can be intimately deposited in an actively respiring medium and can receive adequate illumination. In dye-fed insects this is most likely to occur first in the membranes of the gut wall followed rapidly by implication of other lipoidal membranes (*e.g.*, the neuromuscular sheath) as the highly lipid-soluble dye diffuses throughout the organism. Wherever the integument is sufficiently thin and translucent to allow the passage of visible light, $^1O_2$ is formed *in situ* and rapid, generalized oxidative degeneration of the membranes takes place. Indeed, when contact is sufficiently intimate, involving large specific surface, actual membrane penetration by the dye itself may not be required to produce high lethality, as demonstrated with *Escherichia coli* cells exposed to rose bengal immobilized in polystyrene beads (Bezman *et al.* 1978). Similarly "fixed" and insoluble rose bengal has also been used to sensitize photooxidation of chemical substrates (Logani *et al.* 1982). Thus oxygen, so universally required and intimately perfused throughout every respiring cell, is rendered highly cytotoxic. However, "these reactions have nothing in common with the normal oxygen metabolism of living systems. In fact, they probably represent oxidations of structural components of the cell which do not take place thermally, at ordinary temperatures, since their occurrence would be incompatible with the stability of the cell, but which do take place photochemically, granted the presence of an appropriate photosensitizer" (Blum 1941 a).

c) Insect-dye studies not involving PDA

It would be well at this point to acknowledge a large number of papers reporting extensively on the toxicity to insects of marker dyes and vital stains in which the authors did not discuss the possibility of light sensitization in any depth, if at all. In addition to those included in the review by Barbosa and Peters (1971) there are reports by Daum *et al.* (1969), Hayes and Schechter (1970), Barbosa and Peters (1970), Hendricks (1971), Jones *et al.* (1972), and Bridges *et al.* (1977). Another series of studies has shown that various dyes including eosin, erythrosine, and rose bengal are active inhibitors *in vitro* of several well-established insect detoxication mechanisms including oxidation, hydroxylation, dealkylation, and DDT-dehydrochlorination; these reports, none of which makes reference to PDA, are cited by Jordan and Smith (1981). In a unique application, Matoltsy and Fábián (1946) used larvae of the fruit fly (*Drosophila melanogaster*) as a standard biological indicator to compare the relative PDA activities of several carcinogenic materials; this study, also, was not concerned with either PDA mechanisms of toxicity or with insects *per se*.

III. Studies of PDA in insects[2]

Antonio Barbieri (1928), having considered the reports of PDA on other substrates, proposed the application of this phenomenon in the control of insects, specifically against malaria-bearing mosquitos in Argentina. A number of fluorescent dyes and dye-mixtures in very dilute solution and under various light sources were tested using the larvae of both *Anopheles* and *Culex* spp. The most active toxicants were halogenated fluoresceins and among these a mixture of rose bengal and erythrosine was the most effective suggesting, from today's perspective, a possible synergistic effect. There was high larvicidal activity at a dilution of one part of dye-mixture in 1.5 million parts of water in direct sunlight and no observable harmful effects from the residues. Damage to the larval alimentary tract was noted; tests were made concerning the removal of residual color and on the potability of the water following treatment. Because of rapid bleaching of the dye in direct sunlight—with loss of larvicidal power—an original concentration as high as 1:500,000 was recommended for practical use.

Following upon Barbieri's work, Schildmacher (1950) demonstrated that larvae of three species of mosquitos were highly susceptible to acridine red or rose

[2] To minimize repetition of citations, the bibliography is discussed in chronological order and the scientific names of the species studied are collated in Table III.

Table III. *Insect species studied as substrates for photodynamic action.*

Scientific name	Common name	Investigators
Aedes aegypti (L.)	yellowfever mosquito	Schildmacher (1950)
Aedes triseriatus (Say)	mosquito	Pimprikar *et al.* (1979)
		Carpenter *et al.* (1981)[a]
Agrotis ipsilon (Hufnagel)	black cutworm	Clement *et al.* (1980)
Anopheles maculipennis Mg.	malaria mosquito	Schildmacher (1950)
Anopheles superpictus Grassi.	mosquito	Schildmacher (1950)
Anopheles (unspecified)	mosquito	Barbieri (1928)
Anthonomous grandis		Callaham *et al.* (1975 b)
grandis Boheman	boll weevil	Broome *et al.* (1976)[a]
		Callaham *et al.* (1977 a, b)[a]
Blatta orientalis L.	oriental cockroach[b]	Weaver *et al.* (1976)
Culex quinquefasciatus Say	southern house mosquito	Pimprikar *et al.* (1979)
		Carpenter & Heitz (1980 and 1981 b)
Culex (unspecified)	mosquito	Barbieri (1928)
Diaphania nitidalis (Stoll)	pickleworm	Creighton *et al.* (1980)
Heliothis zea (Boddie)	corn earworm	Creighton *et al.* (1980)
	(bollworm; tomato fruitworm)	
Hermetia illucens (L.)	black soldier fly	Pimprikar *et al.* (1980 a)
Musca autumnalis De Geer	face fly	Fondren & Heitz (1978 a)
		Fairbrother (1978)
		Fairbrother *et al.* (1981)
Musca domestica L.	house fly[b]	Yoho *et al.* (1971, 1973, 1976)
		Fondren *et al.* (1978)
		Fondren & Heitz (1978 b and 1979)
		Pimprikar *et al.* (1980 a and b)
		Respicio & Heitz (1981)
		Sakurai & Heitz (1982)

Periplaneta americana (L.)	American cockroach	Weaver *et al.* (1976 and 1982)[a]
Pieris brassicae (L.)	large white cabbage butterfly	Lavialle & Dumortier (1978)
Solenopsis richteri Forel	black imported fire ant[b]	Broome *et al.* (1975)
		Callaham *et al.* (1975 a)[a]
Tenebrio molitor L.	yellow mealworm	Graham *et al.* (1972)
Trichoplusia ni (Hübner)	cabbage looper	Creighton *et al.* (1980)

[a] Report contains biochemical analyses in addition to mortality data.
[b] See "Note added in proof," p. 99.

bengal at 1:100,000 dilution with illumination, even when removed to pure water following dye-ingestion. The suggested cause of death was photochemical reaction in the midgut wall involving those wavelengths characteristic of the dye absorption spectrum. Various indoor and outdoor tests proved the dye to be an effective larvicide, without observed harm to the fish (*Gambusia* sp.) which were present.

The modern era of photodynamic insecticide studies began with the report by Yoho *et al.* (1971) presenting mortality vs. concentration data on dye-fed adult house flies under various lighting regimes. The results suggested that further studies were fully warranted on mode-of-action and on practical application of dyes of the xanthene group to insect control. Graham *et al.* (1972 b) described the toxic effect of methylene blue, both dietary and injected, on illuminated yellow mealworms with the suggestion that this species would be a suitable test organism for comparative bioassays of photosensitizing chemicals in general. Natural solar intensity was shown to be greatly in excess (up to 25X) of that required for effective kills with this dye.

The effect of six fluorescent dyes on mortality of adult house flies in the dark, and in natural and artificial light was systematically explored by Yoho *et al.* (1973). They confirmed a graded effectiveness among the xanthene dyes rose bengal (100% mortality), erythrosine (96%), eosin Y (49%), and they explained the relatively low toxicity of methylene blue (8%) in terms of a lack of coincidence of its absorption spectrum with the spectral bandwidth of the light supplied. These authors also described the visible acute symptoms displayed by the poisoned insects and they correlated dye concentration, light intensity and the length of exposure-time with the mortality data.

A possible implication of the neurotransmitter control enzyme acetylcholinesterase (AcChE) was suggested by Broome *et al.* (1975) after studying behavioural symptoms of rose bengal-poisoned fire ants. These authors, also, showed the toxicity to be dependent on dye concentration, on incubation period in contact with the dye, on continuous illumination, light intensity and exposure-time. The photooxidation type of toxicity was thus confirmed to be light-dependent but the presence of a separate, much slower, *light-independent* toxic mechanism was also discovered in the fire ant. The use of baits treated with rose bengal was proposed for control of fire ants both in light as well as in the dark nest to exploit this non-PDA process.

Callaham *et al.* (1975 a) demonstrated that the AcChE in homogenates of whole heads from the fire ant was inhibited *in vitro* by low levels ($\sim 10^{-7}$ *M*) of several halogenated xanthene dyes but the levels of the same enzyme in similar homogenates made from ants previously killed by the dyes were not depressed from normal. Rhodamine B (a nonhalogenated xanthene dye) was found to be noninhibitory *in vitro,* quite the opposite to rose bengal, phloxine B, erythrosine, eosin Y, and fluorescein. Correlations were observed between dye structure (degree and weight of halogenation in the fluorescein ring system), phosphorescence quantum yield, and the PDA effectiveness. A quantitative comparison of dye activity was accomplished by calculating rate-constants of inactivation of AcChE under standard illumination. Oxygen consumption was also quantitated, confirming all of the requirements of the definition of PDA. No specific site of toxic action of PDA was located.

The development of quantitative methodology for comparing the effectiveness of various dyes under differing light conditions was continued by Callaham *et al.* (1975 b) using the adult boll weevil as substrate; reaction rate-constants for specific toxic effect were related to dietary content and to tissue levels of rose bengal. Acute physical symptoms were described. Addition of α-tocopherol, a known anti-oxidant, had no apparent effect on the PDA of rose bengal in the weevil.

Food, drug, and cosmetic dye additives were screened for PDA toxicity using adult house flies (Yoho *et al.* 1976). Nine of the 14 dyes tested produced up to 100% mortality in illuminated flies, particularly in those receiving liquid diets. The possibility of hazard to the public from such photosensitizing chemical additives was suggested. In two species of cockroach, Weaver *et al.* (1976) observed hemolymph volumes to decrease significantly following treatment with rose bengal or erythrosine and subsequent illumination. The hemocoel fluids were seen to transfer rapidly from the body cavity into the alimentary canal with a large increase particularly in crop volume. Changes in membrane permeability were suggested because of altered potassium levels in the hemolymph.

Broome *et al.* (1976) attempted to elucidate the slow, light-independent reaction of rose bengal on the boll weevil although, by definition, this is not the result of PDA. There were large deviations from controls in weight, levels of water, lipids, protein, and amino acids which suggested a chemically-induced lethal energy stress in the organism. These findings were quantitated and related to time-elapse (following dye ingestion) by Callaham *et al.* (1977 a) who again postulated a cumulative energy stress mechanism unrelated to light-sensitized PDA. High rates of PDA-inhibition of boll weevil enzymes lactic dehydrogenase (LDH) and AcChE were measured quantitatively *in vitro* for several xanthene dyes although, again, no change in the *in vivo* levels of these enzymes in poisoned weevils could be demonstrated (Callaham *et al.* 1977 b). As before, the efficiencies of photooxidation *in vitro* were seen to correlate with dye structure (number and mass of halogen substituent atoms in the dye molecule) and its effectiveness in exciting oxygen to the singlet state. The photochemical mechanism for dye-sensitization of oxygen was schematically illustrated, but "the *in vivo* target of dye-sensitized photooxidation remains as yet undescribed."

Relative toxicities of six xanthene dyes towards house flies were quantitated by determining rate-constants of photooxidation and relating these to actual tissue levels of the dyes (Fondren *et al.* 1978 a). Correlation was suggested but not confirmed between the effectiveness of the dyes and the number and mass of halogen substituents and resulting phosphorescence quantum yields. There were indications of insect species-variability in reaction to the dyes which was not related to dye structure, as well as some evidence of feeding inhibition at high dietary levels of dye.

Fondren and Heitz (1978 b) examined the PDA effectiveness of eight xanthene dyes on the adult face fly and reported a graded response paralleling the mass of halogenation in the dye molecule (*i.e.*, rose bengal > erythrosine > phloxine B > eosin Y > tetrachlorofluorescein > fluorescein). In the light-independent reaction, also, a significant degree of species variation became apparent on comparing reactions to the dyes, the house fly and face fly being much more susceptible to rose bengal than the boll weevil or fire ant. The non-halogenated rhodamine B was surprisingly highly toxic to the face fly in the absence of light. Among the halogenated xanthenes toxicity in the light-independent reaction was suggested to be due to an "organochlorine-type" of poisoning since obviously it is not due to PDA.

The critical importance of intensity of light in the PDA effectiveness of six xanthene dyes against the house fly was demonstrated by careful calculation of more-comprehensive photooxidation rate-constants and by photon counting (Fondren and Heitz 1978 c). It was proposed that more "photons per fly" are required to attain a 50% mortality level at lower flight fluxes because of the insects' ongoing repair and recovery processes whereas these are overwhelmed when higher light intensities are used. Mortality data were reported (Lavialle and Dumortier 1978) on methylene blue-fed larvae of the large white cabbage butterfly exposed to fluorescent light. In this species, as with others, mortality is proportional to dietary level of the dye and is dependent on parameters of photo-irradiation such as light intensity, duration, and wavelength. The authors demonstrated adequate transmission of wavelengths required for sensitization with methylene blue (500 to 700 nm) through the leaves of cabbage, corn, and endive on a typical sunny day.

Heitz and Wilson (1978) addressed the crucial subject of environmental degradation of the dyes and toxicity of their terminal residues. Rose bengal was shown to degrade in sunlight; its characteristic absorption spectrum disappeared concurrently with the appearance of new spectra indicative of the formation of intermediate or terminal residues which were not identified. The photodegraded dye solutions were not toxic to house flies either in the light (by PDA) or in the light-independent reaction (non-PDA), nor did they retard the growth of two species of bacteria, both of which were inhibited by nonphotodegraded rose bengal. Comparative data–rate constants for photodegradation–for six dyes were determined and a relationship appeared to exist linking position, as well as mass, of halogen substituents to susceptibility to photodegradation. Erythrosine was the most rapidly degraded under conditions in which fluorescein and rhodamine B were stable. Rapid disappearance from the environment of the photodynamic toxicity of erythrosine and rose bengal and their replacement by nontoxic residues was implied.

The effectiveness and efficiency of various types of light source in producing dye-sensitized toxicity were quantitatively evaluated by Fondren and Heitz (1979) using house flies, six xanthene dyes, and three different light sources. Sunlight was the most effective followed by sunlamp and then by fluorescent light but the latter was 2.5 to 3 times more energy-efficient than the other two sources. Light-independent toxicity also was measured, with rose bengal proving to be twice as effective against house flies as was erythrosine; a need to account for the contribution of this non-PDA process to total effectiveness of the dyes was discussed.

Rose bengal was re-examined as a photodynamic toxicant towards mosquito larvae (2 spp., see Table III) by Pimprikar *et al.* (1979) in the continuing search for alternatives to the resistance-prone organophosphates and carbamates. At very low concentrations of the dye in water (1 to 20 ppm) with fluorescent lamp illumination *Culex* larvae were found more susceptible than *Aedes* larvae and, in general, earlier instar larvae were more rapidly killed than 4-th instar forms. Rate-constants were calculated. Physiological and morphological abnormalities, manifested by difficulty to molt and separate from old cuticle, were reported. At these low dye-concentrations and rapid, acute PDA mortalities, no light-independent toxicity to mosquitos was observed.

The toxicity of dietary rose bengal to larvae of the cabbage looper, the corn earworm, and the pickleworm was compared by Creighton *et al.* (1980) who reported a relatively high degree of resistance to photodynamic poisoning in the earworm under the experimental conditions used. In order to achieve a mortality rate comparable to the other two species, so much dye was required that 68% of the nonilluminated, dye-fed earworm controls were succumbing to the light-independent toxic mechanism.

The first modern attempt at practical application of a photodynamic insecticide was the use of erythrosine sprays for fly control in chicken houses by Pimprikar *et al.* (1980 a). Using weekly spraying on the manure over a four-week period, house fly populations were reduced by about 90% with little or no apparent harm to the beneficial soldier fly. The dye, as such, was not persistent in the manure, only 20% being recoverable after one week. The only light source was indirect sunlight. In a technical news item Heitz (1979) discussed the advantages, cost, and safety of this application.

A light-dependent but latent toxicity mechanism of rose bengal to mosquito larvae was reported by Carpenter and Heitz (1980); at 1 to 30 ppm of dye in water, illumination caused not only a high acute mortality but also it initiated a latent lethal process among survivors removed early from the light. At these dye concentrations, no light-independent mortality was observed. First instar larvae were much more suceptible than 4-th instars, 13:1 for acute toxicity and 42:1 for the latent mechanism. The gut tract appeared to be destroyed. The added contribution of this latent toxicity increased the overall mortality level to 10 times the acute PDA toxicity alone. The use of rose bengal was described as "especially promising . . . under conditions when mosquito control is necessary but application of more toxic materials would not be feasible."

Laboratory studies of the effectiveness of rose bengal, erythrosine and phloxine B against larvae of the black cutworm were described by Clement *et al.* (1980), seeking an alternative to conventional pesticides. Only rose bengal was highly effective in this study and only at the highest light intensity (≅ 12% of sunlight for 4 hr followed by incubation in the dark). The difficulty of controlling a light-avoiding species such as the cutworm by photodynamic means was acknowledged.

Pimprikar *et al.* (1980 b) demonstrated that all life stages of the house fly are susceptible to some form of xanthene dye poisoning, including acute and delayed mortality, under minimal light conditions. Adults deposited fewer and less viable eggs, treated eggs were less likely to hatch, and larvae exhibited up to 81% reduction in ultimate adult emergence. This work supports a growing body of evidence that a number of toxic mechanisms, in addition to PDA, are at work in the total insecticidal activity of the xanthene dyes and that the relative contributions of each mechanism would be difficult to define.

A bioassay procedure using face flies was developed (Fairbrother *et al.* 1981) to monitor insect mortality through three successive life stages, from dyes artificially incorporated into manure from steers. Larval, pupal, and adult stages were affected, on illumination, to a degree depending on dye concentration. With rose bengal and erythrosine there were high mortalities in larvae, decreased pupation, low emergence rates, morphological abnormalities, and lowered adult viability. Cattle fed encapsulated erythrosine produced manure which was similarly toxic, in proportion to the dye levels present. Additional findings pointing to interspecies differences in relative effectiveness of the various xanthene dyes were discussed.

A marked synergistic action by fluorescein on photosensitization by rose bengal was observed and quantitated by Carpenter *et al.* (1981 a) using *Aedes* mosquito larvae as the test organism. The proposed mechanism for this enhancement of the PDA toxicity of a very active dye by its nonhalogenated and relatively inactive analogue was based upon the overlap of their fluorescence and excitation spectra. (This phenomenon would also explain the qualitative observation of Barbieri [1928; see above] concerning the unexpected efficacy of mixtures of rose bengal and erythrosine [author].) The same degree of synergism was seen to apply *in vitro* to the rose bengal-sensitized photoinhibition of the enzyme glyceraldehyde-3-phosphate dehydrogenase.

Erythrosine was closely examined by Carpenter and Heitz (1981 b) for its effect on larvae of the *Culex* mosquito. They established that both the acute and the latent light-dependent toxicities as well as the slow light-independent mechanism all occur in this species using erythrosine, just as they do with rose bengal. The contributions of these various modes-of-action to total mortality were compared, although the rapid photodegradation of erythrosine makes its PDA component difficult to assess. After a consideration of environmental impact, the authors suggest that erythrosine has potential for use in mosquito control under conditions unfavourable to the employment of more toxic insecticides.

Rhodamine dyes are nonhalogenated members of the xanthene group and are highly fluorescent rather than phosphorescent; theoretically, their toxicity should therefore be less light-dependent than that of "the singlet oxygen generators" such as eosin Y, phloxine B, erythrosine, and rose bengal. This was confirmed in adult house flies by Respicio and Heitz (1981), who also measured the differential toxicity of rhodamine B and rhodamine 6G under various conditions (feeding vs. injection, light vs. dark, varied light exposures, dietary level vs. tissue level of dye, and effectiveness against males compared to female flies).

Effects of erythrosine on hematocrit levels in the American cockroach were reported by Weaver *et al.* (1982). The photodynamic mechanism resulted in diminishing numbers of hemocytes, whereas the dark, light-independent reaction caused higher levels except at excessively high injected doses of dye where there was a 25% decrease in cell count. Sakurai and Heitz (1982) have studied the carryover and enhancement of photooxidative toxicity in house fly adults exposed to light after sub-lethal quantities of the dye had been administered at an earlier life stage. Rose bengal and erythrosine both caused inhibition of pupation and decreased pupal weight when larvae were kept in the dark. After comparing the effects of ingested vs. injected dyes, the authors suggest that "the target organ of the dye-sensitized photooxidation reaction may not be the alimentary canal . . . [and] is as yet unknown."

IV. Discussion

The studies of PDA toxication in insects indicate that certain photosensitizing dyes, the halogenated fluoresceins in particular, possess potential as insecticides and could offer a number of immediately obvious advantages. Whether these can outweigh some of the equally-evident, as well as some of the more subtle disadvantages and dangers, remains in some doubt. Superficially it would seem that erythrosine, at least, because of its use for nearly a century as a permanently-listed food colorant could be used quite safely in the environment at the relatively low concentrations required for an effective photodynamic insecticide. However, these are synthetic "coal-tar dyes," some of which are demonstrably harmful; for the present, erythrosine must be "suspect-by-association" and careful selection of sites and applications will be required, as discussed below.

a) General considerations

1. Advantages.–

(1) Powered by daylight and fueled by atmospheric oxygen, the relatively inexpensive dyes complete an insecticidal system which is economically attractive.

(2) The most effective of the dyes, rose bengal and erythrosine, are water-soluble salts and easily applied to farm water ponds and liquid manure pools, for fly control in manure pits around stables, poultry houses, feed-lots, and swine barns, and to marshes and ditches where mosquito larvae abound. They do not require the adjuvents (synergists, emulsifiers, oil solvents, spreaders, etc.) which are common to conventional formulations.

(3) Given adequate light–and normal diffuse daylight is often sufficient–the dyes act rapidly on the larval and adult forms of many species of insects to effect a significant reduction in overall population.

(4) When used in liquid diet or wet media, high concentrations of the dye are not required because of intimate contact and rapid ingestion by the target organism.

(5) Photodynamic activity, itself, appears to degrade rapidly in light and is lacking in the terminal residues remaining in the medium.

(6) When used in the manner which is insecticidally optimal, none of the halogenated fluoresceins appear to exert any *photodynamic* hazard to domestic animals or to man.

(7) Photodynamic toxicity is a multi-component system (Table I) and exists only when all the requisite factors are brought together. The material therefore does not present the immediate lethal hazard which accompanies storage of large quantities of conventional insecticides.

2. **Disadvantages and hazards.–**

(1) One of the prime objectives of modern integrated pest control strategy is a reduction in the number of chemicals introduced into the environment. The use of photosensitizing dyes would not assist in attainment of this goal.

(2) By the very nature of its mode-of-action, PDA toxicity, in itself, is not likely to be species-specific. Any insect or other small organism, harmful or beneficial, which ingests and generally distributes the unchanged dye, possesses a translucent integument, does not avoid the light, and carries on normal aerobic metabolism, will be vulnerable to singlet oxygen attack. The development of resistance by any species would have to be through modification, directly or indirectly, of one of these factors.

(3) Obviously, photodynamic action cannot be effective against insects which remain in the soil during daylight hours or which engage in light-avoidance behaviour such as leaf-rolling, burrowing under bark, etc. Similarly, the dyes are ineffective as PDA toxicants when applied in dark storage areas.

(4) Several factors militate against the use of PDA insecticides on field crops. Photodynamic action is also harmful to plant-life, according to conditions prevailing, and the possibility of phytotoxicity would have to be considered for each type of crop. Rapid photodegradation of the dye would necessitate frequently-repeated applications and its water-solubility would be a disadvantage in the event of rainfall following application. Insect kill is not obtained on contact and considerable host-plant damage must be tolerated in order to attain adequate larval ingestion and internal distribution of the dye. A higher level of dye must be incorporated in a solid diet in order to ensure sufficiently high tissue levels for effectiveness. Finally, unsightly staining with garish fluorescent colors is not conducive to general crop use.

(5) Subtle hazards of the dyes

It is already evident that the photosensitizing dyes cannot be applied for every purpose or in a widespread manner. Urgent attention of the medical profession has been called to the health hazards of neglecting the phototoxicity factor when using these light-absorbing substances (Valenzeno and Pooler 1979). In considering the hazards to nontarget organisms when using such materials as insecticides, it is essential to distinguish clearly between photooxidative injury (PDA) and the chemical toxicity of the parent dye in its original form, or as a metabolic intermediate or, finally, as a terminal residue. Because of selected sites of application, of rapid loss of PDA toxicity through photoinactivation, and of the improbability of the dye to concentrate at any photo-vulnerable site in larger animals, there will be little hazard, if any, to domestic animals or to man due to PDA, but the other mechanisms of toxicity must be considered at greater length. There can be no justification for ignoring the warning signs which have been associated with the rhodamines, phloxine B, eosin Y, rose bengal, and, to a lesser extent, with erythrosine.

3. The parent dyes.–The majority of these dyes are not feasible candidates for practical insecticidal use. Fluorescein, uranine, and eosin Y exhibit relatively low PDA effectiveness in insects. Methylene blue, also not as active in natural light as the halogenated fluoresceins, is decidedly photomutagenic through DNA degradation in the *Salmonella typhimurium* (Ames) test (Gutter *et al.* 1977, Chung *et al.* 1981) and both the rhodamines, B and 6G, are now registered as being carcinogenic to rats (IARC 1978).

Within the past five years there has been an increasing number of disturbing reports concerning the xanthene dyes with regard to subtle immunotoxic, mutagenic (genotoxic, clastogenic), carcinogenic, and enzyme-inhibitory effects. This extensive and often conflicting literature cannot be reviewed here but, for perspective, a few illustrative examples are in order. The phloxines have been shown to be mutagenic to *S. typhimurium* and to *Bacillus subtilis* when photoactivated although they were not so in the absence of light; they were growth-inhibitory or cytotoxic in tests with *Escherichia coli* and were reported to be mutagenic to cultured human diploid cells and to induce DNA strand breakage in hamster ovary cells even in the absence of light (Maus *et al.* 1981). Chromosome breakage by both halogenated and non-halogenated xanthenes (rose bengal and rhodamine B) was seen in varied cell systems (Au and Hsu 1979, Lewis *et al.* 1981), immunomodulatory activities by rhodamine B were described by Wess and Archer (1982) and DNA damage by phloxine, and possibly erythrosine, was discussed by Haveland-Smith and Combes (1980). The DNA polymerase I of *E. coli* is potently and irreversibly inhibited by photoactivated rose bengal and to only a slightly lesser extent by erythrosine (Stern *et al.* 1980). The possible significance of these findings was amplified by McCann and Ames (1976) who have concluded that many chemical carcinogens probably initiate cancer through damage to DNA and who have shown that almost all carcinogens (90%) are also mutagens. This high correlation of mutagenicity with carcinogenicity was independently confirmed (86%) by Sugimura *et al.* (1976) who stressed the importance of identifying mutagens as a screening procedure for likely carcinogens. Erythrosine was assessed as "probably a toxic substance but not a carcinogen" using rat embryo cells *in vitro* and hamsters *in vivo* (Price *et al.* 1978). The degree of influence exerted by ambient light was not a parameter in many of these *in vitro* cytological investigations and, therefore, the results often appear contradictory. The inconclusive nature of these studies has been reviewed extensively by Haveland-Smith *et al.* (1981).

The observations of Tonogai *et al.* (1978, 1979 a and b) on the toxicity of the xanthene dyes are more conclusive and they raise doubts concerning the safety of their use in water pools containing fish. The dyes themselves (at concentrations of 1:10,000) were acutely toxic to carp (unspecified) and to "himedaka" (Medaka: *Oryzias latipes*) through binding to the protein of gill tissue and subsequent asphyxiation; this was particularly true of the higher halogen-content dyes rose bengal, erythrosine, and phloxine. The xanthene dyes, as a class, were much more toxic to fish than the other dye-types tested (azo and triphenylmethane dyes). Rose bengal and phloxine were also mutagenic in rec.-assay tests with *B. subtilis*; rose bengal, phloxine, and erythrosine all inhibited the enzymatic activity of trypsin by 40 to 50%, as well as binding strongly to albumin in direct proportion to the degree of their toxicity to fish. Sako *et al.* (1980) have reported that the xanthene dyes, especially phloxine, were more cytotoxic to fetal rat hepatocytes than any of the other classes of food colorants tested. Eleven food, drug, and cosmetic xanthene dyes, including all of the halogenated fluoresceins discussed above, were shown to alter the electrophysiological properties of molluscan neurons by increasing the relative potassium permeability of the membranes and their activities were highly correlative with their lipid solubility (Levitan 1977); it follows that the "best" photodynamic insecticides [*i.e.*, the highly halogenated rose bengal, erythrosine, and phloxine B which are the most soluble in lipo-membranes (Valenzeno and Pooler 1982)] exhibited the highest disruptive activities.

These are only a few of the available examples of apparent chemical biotoxicity *in vitro* of the xanthene dye molecules. Their significance in terms of potential hazard to man when used at low insecticidal levels in the "real" environment cannot be assessed at present but, obviously, some caution is required with regard to natural ecological systems.

4. Metabolites and terminal residues.–We have insufficient knowledge-in-depth concerning metabolites and residues, either as to what they are or concerning their chemical toxicity (as opposed to photodynamic activity of the original parent dye). When a halogenated fluorescein dye photodegrades it changes its absorption spectrum, bleaches, and loses its photodynamic activity, but the products nevertheless will contain halogens and halogenated aromatic materials whose structure and toxicity must remain of prime concern. Most recently, Valenzeno and Pooler (1982) have suggested that the bleaching of these dyes involves a change in the fluorescein skeleton itself and is more than merely a conversion from one derivative to another by dehalogenation; the unidentified bleached residues appear to be PDA-inert (in photohemolysis of erythrocytes) but nothing is known of their chemical toxicity. Carpenter and Heitz (1981 b) have confirmed the rapid disappearance of erythrosine from illuminated aqueous solutions (similar to the photodegradation and loss of photodynamic activity of rose bengal described by Heitz and Wilson in 1978) but there were no residue identification or mammalian toxicological studies on the photodegradation products; in both cases, photodegradation was carried out in a manner which allowed volatiles (including halogens) to escape the system.

In canned fruits colored by erythrosine, electrochemical reduction is known to produce free fluorescein and crystalline iodine (Dickinson and Raven 1962). Tonogai *et al.* (1979 a) have reported that both rose bengal and erythrosine were deiodinated by photodecomposition (low-pressure mercury vapour lamp), that the principle nonvolatile residues from these two dyes (in 80 to 90% yield) were tetrachlorofluorescein and fluorescein, respectively, and that these two materials were less toxic to fish than their parent dyes only if isolated and tested separately. On the other hand, the unresolved residual mixtures following illumination were more toxic than the parent dye solutions and it was the liberated iodine which accounted for the increased toxicity to fish of the photodegraded mixture. These workers (1979 b) also demonstrated that the liberated iodine, to a greater extent than free bromine, was significantly mutagenic in rec.-assay tests with *B. subtilis*. The previously assumed relative safety of erythrosine (tetraiodofluorescein) is thus brought into question since it is known to photodegrade particularly rapidly in sunlight and undoubtedly releases iodine in the same way. These studies made use of dye-concentrations not unlike those which would be found in any practical application of the dyes as insecticides.

5. Light-independent and latent effects.–Closely related to the chemical toxicity of metabolites and terminal residues are the unsolved questions of the light-independent toxic mechanism and the light-dependent but latent process. Although these two modes-of-action are additional to PDA and increase the ultimate effectiveness of the dyes as insecticides, they are complicating and unwelcome factors when attempting to assess environmental hazard. These as-yet poorly understood mechanisms may be a composite result of the many mutagenic, inhibitory, and other biochemical disturbances discussed earlier. The "dark action" toxicity of the xanthene dyes, of rose bengal in particular, had already been investigated by Blum (1941 b) and shown to be based upon a mechanism quite different from that of PDA.

At high dye concentrations there is a light-independent mortality in which dye-fed insects, maintained entirely in the dark, will slowly die over 10 to 12 days. This effect was overlooked in the early experiments using illumination where high acute mortalities were encountered in 1 to 2 days and the dye-treated, dark controls were discarded prematurely. In 1975 Broome *et al.*, using the imported fire ant, described an 8-to-9 day period for 50% mortality with rose bengal in the dark while similarly-treated insects in the light suffered 50% mortality in only 24 hr. At first considered to be peculiar to the fire ant, this light-independent process has since been encountered in the boll weevil (Broome *et al.* 1976, Callaham *et al.* 1977 a), in the face fly (Fondren and Heitz 1978 b), in the adult house fly (Fondren and Heitz 1979), in all life-stages of the house fly (Pimprikar *et al.* 1980 b, Sakurai and Heitz (1982), in the corn earworm (Creighton *et al.* 1980) and in mosquito (*Culex*) larvae (Carpenter and Heitz 1981 b); all of the xanthene dyes have been implicated to some extent. Since this mode of toxicity is not due to PDA and was first noted in association with the halogenated xanthene dyes it was suggested to be due primarily to a type of organohalogen chemical toxicity (Fondren and Heitz 1978 b) which results in the symptoms of energy stress proposed by Callaham *et al.* (1977 a). Other studies, however, the most recent by Respicio and Heitz (1981), re-confirm that the nonhalogenated xanthenes (*e.g.*, rhodamines B and 6G) also possess high light-independent toxicity. The inhibition by xanthene dyes of several well-known insect detoxication systems (Jordan and Smith 1981) may play a role in this light-independent process. Similar enzymatic disruptions in chitin formation and maturation may also account for the frequently reported morphological abnormalities and molting difficulties associated with these dyes.

The light-dependent latent toxic process was seen in dye-treated pre-imaginal mosquito larvae which were maintained in the dark after being illuminated too briefly for acute PDA mortality to occur; the ensuing death takes place too soon and at too low a dye concentration to be attributable to the light-independent reaction. It was observed when using very low concentrations of rose bengal or erythrosine (Carpenter and Heitz 1980, 1981 b), and several days elapsed before the pre-illuminated insects finally died in the dark; dye-treated controls, maintained entirely in the dark, were not affected. There appear to be species differences in the response to interrupted illumination in that Broome *et al.* (1975) found that continuous light exposure was required to kill the fire ant and premature removal from the light resulted in recovery. Also, either the repair and recovery mechanisms (as reported in house flies subjected to low intensity illumination by Fondren and Heitz (1978 b)) are not operative in the mosquitos studied, or there is more than one toxic mechanism at work.

Clearly, until all of these more subtle processes are thoroughly understood, practical use of the dyes should be carefully limited.

Finally, there have been intimations that the PDA mechanism may not of itself be adequate to provide practical insect control but when enhanced by the accompanying light-independent and latent processes the overall effectiveness could suffice for some specific applications (Pimprikar *et al.* 1980 b, Fairbrother *et al.* 1981). It seems dangerous to resort to these insidious and little-understood mechanisms, if such is required, to rescue a borderline technology. Although the slower-acting, chemically toxic processes must be thoroughly assessed as hazardous side-effects, particularly to nontarget organisms, the use of the photodynamic effect for insecticidal purposes should be seen to succeed or fail on its own merits.

b) A special case for erythrosine?

Of all the dyes which have been investigated for PDA toxicity to insects, erythrosine alone is considered sufficiently safe to be universally permitted as a food colorant. In contrast, rose bengal is listed for food use in only two countries (Noonan and Meggos 1980); also, in addition to the negative factors already cited, this dye contains the potential for retaining toxic chlorinated hydrocarbon structures among its residues following de-iodination and its toxicology is not as well-studied as that of erythrosine. The latter, as "FD&C Red No. 3," has been used in over 40 countries and for more than 80 years without exhibiting any proven harmful effect in the general population; approximately 50 tons of this dye were used in foods, drugs, and cosmetics in the United States in 1967 (derived from Certified Color Industry Committee 1968). In Canada, where maximum permissible levels are legislated, erythrosine is permitted in foods, singly or in combination with other specified dyes, to a total of 300 ppm. The acceptable daily intake (ADI) for human beings is 0 to 2.5 mg/kg, suggesting that whatever its borderline or suspected faults, erythrosine is the least hazardous of the xanthene dyes and a great deal safer than most of the conventional pesticides, none of which could be tolerated as a food additive. Good methods are available for assessment of its quality (Marmion 1979, Goldberg and Calvey 1982) and for its purification (Peeples and Heitz 1981).

The acute *oral* LD_{50} of erythrosine in rats and mice is high (6.7 to 7.4 g/kg) and its long-term "no-untoward-effect" level is 0.25% in the diet, equivalent to 160 to 170 mg/kg/day over 10% of life span (Butterworth *et al.* 1976 a). In pigs no such threshold could be established, possibly due to the brevity (2% of life span) of the experiment (Butterworth *et al.* 1976 b). Even following intravenous injection in mice, the LD_{50} for erythrosine is relatively high: 700 mg/kg, compared to 550 for eosin Y, 310 for phloxine B, and very low for rhodamine B (Lutty 1978). Rats appeared to metabolize about 30% of the dye in the tissues, excreting the unchanged remainder in feces; the urine contained no erythrosine or any of the conjugates usually found in the metabolism of other dyes (Daniel 1962).

Because of its highly labile iodine content, studies of the effects of dietary erythrosine on the thyroid gland have been made in four mammalian species. In both rats and mice there was no detectable effect on the thyroid (Butterworth *et al.* 1976 a), but in gerbils (Collins and Long 1976), and in pigs (Butterworth *et al.* 1976 b), there was measurable interference with normal thyroid metabolism accompanied by enlargement and weight-increase of the gland. The magnitude of these effects, in relation to dose, does not suggest that erythrosine would be unsafe in the open environment when applied at insecticidal levels and subject, as it is, to rapid photodegradation with loss of iodine.

Lutty (1978) has compared the toxicity and routes of elimination from mice of erythrosine and other fluorescent dyes. Fairbrother (1978) observed some disturbing effects on ruminant digestive processes, *in vitro*, which require further study. In an exchange of letters (Kon-Heitz 1979), Heitz has enlarged upon his views concerning the safety of erythrosine towards mammals as compared to insects and these have been restated by Carpenter and Heitz (1981 b).

There may be a challenge concerning the use of erythrosine near the sites of food production on the grounds of its possible implication, as a food dye, in the hyperkinesis-learning disability syndrome (HLD) in children (Feingold 1975,

Swanson and Kinsbourne 1980 a). Aside from the fact that the overall HLD hypothesis remains controversial and un-proven (Conners 1980, U.S. National Institutes of Health 1982, Anonymous 1982), recent evidence is strongly indicative that erythrosine is not involved (Swanson and Logan 1980 b, Goldenring *et al.* 1981). In any case, there is little possibility that the dye, used at low insecticidal concentrations, could gain entrance into the food chain either undetected or in amounts approaching those which are deliberately incorporated into food every day.

It appears that unchanged erythrosine, at low levels, is relatively innocuous to larger animals either through PDA or chemical poisoning. Its effect, and that of its terminal residues, on other fauna and flora at sites of insecticidal application require further investigation. Meanwhile, erythrosine remains the preferred candidate for developmental tests of xanthene dyes as photodynamic insecticides.

Summary

There is ample evidence that phosphorescent dyes, particularly the halogenated fluoresceins of the xanthene group, possess insecticidal activity when applied to insects as a dietary additive. The possibilities of practical exploitation of this phenomenon appear to be marginal but are still under investigation. Both light-dependent photodynamic action (PDA) and light-independent (chemical) toxic mechanisms are involved. The basis of PDA toxicity is the dye-sensitized excitation of molecular oxygen to a highly-reactive singlet state which causes acute lethal effects, probably through photooxidative disruption of cell membranes. The mode-of-action of the slower light-independent toxicity appears to be a composite of a number of biochemical disturbances through mechanisms quite separate from that of PDA. No studies have been reported concerning the possible development of resistance in insects towards photodynamic action and any selectivity which has been seen may be due to physical factors such as attenuation or exclusion of light or low level of dye ingestion.

Rose bengal and erythrosine are the only dyes which presently appear to warrant further investigation as insecticides; others (*e.g.*, eosin Y, fluorescein, uranine, phloxine B, rhodamine B) are either too PDA-inactive or too chemically toxic for practical use. Rose bengal is generally more toxic to insects and more resistant to rapid photoinactivation than erythrosine but the latter material appears to be less chemically toxic and to present the least environmental hazard for further applications testing. Neither of these materials, following application at efficient insecticidal levels, presents any significant photodynamic hazard to man or to domestic animals.

A complete description of the chemical toxicity of the halogenated fluoresceins and the nature and toxicology of their terminal residues has not been established but there is much evidence that most of these dyes are mutagenic or otherwise cytotoxic in a number of standard *in vitro* tests. Erythrosine has been the least implicated by such tests but must remain suspect-by-association. It is unclear, in many cases, whether the active mutagens are the parent dyes operating through their own PDA and/or chemical mechanisms, or are metabolites, or degradation products; in experiments with fish it was the photo-liberated halogens, especially iodine, which proved to be highly toxic *in vivo* as well as being mutagenic to bacteria and enzyme-inhibitory *in vitro*. The significance of these findings in actual outdoor insecticidal applications requires further study.

Present indications are that erythrosine, at low concentrations and in well-lighted areas, could be safely and effectively used in a few selected applications where there is reason to avoid use of conventional insecticides of much higher acute toxicity to domestic animals. Examples would be against fly populations in manure pits and ponds around livestock barns, feed-lots, poultry operations, and other sites of food production; similarly, mosquito larvae could be attacked in nonshaded water ponds, marshes, and poorly drained ditches.

Current studies of photodynamic toxication in insects contribute significantly to the elucidation of mechanisms of photosensitization and other interactions of light with xenobiotic and natural sensitizing materials and will lead to a better understanding of the resultant effects on ecological and evolutionary processes. However, until more is known concerning the chemical nature and long-term toxicological effects of these materials and their terminal residues, there appears to be little likelihood of large-scale acceptance of the photosensitizing dyes as insecticides.

References

Amagasa, J.: Dye binding and photodynamic action. Photochem. Photobiol. **33**, 947 (1981).

Anonymous: Diet-hyperactivity link remains unconfirmed. Chem. Eng. News **60** (4) 11 (1982).

Au, W., and T. C. Hsu: Studies on the clastogenic effects of biologic stains and dyes. Environ. Mutagen. **1**, 27 (1979).

Barbieri, A.: Sensibilizadores fluorescentes como larvicidas. Acción fotodinámica de la luz. Riv. Malariol. **7**, 456 (1928).

Barbosa, P., and T. M. Peters: Retardation of growth-rate in *Aedes aegypti* (L.) larvae exposed to vital dyes. J. Med. Entomol. **7**, 693 (1970).

——— ——— The effects of vital dyes on living organisms with special reference to methylene blue and neutral red. Histochem. J. **3**, 71 (1971).

Barratt, M. D., J. C. Evans, C. A. Lewis, and C. C. Rowlands: Comparison of the photodynamic action of rose bengal and tetrachlorosalicylanilide on isolated porcine erythrocyte membranes. Chem. Biol. Interact. **38**, 215 (1982).

Bezman, S. A., P. A. Burtis, T. P. J. Izod, and M. A. Thayer: Photodynamic inactivation of *E. coli* by rose bengal immobilized on polystyrene beads. Photochem. Photobiol. **28**, 325 (1978).

Blum, H. F.: Photodynamic action and diseases caused by light. Amer. Chem. Soc. Monograph series No. 85. (a, p. 4; b, p. 83) New York: Reinhold (1941).

Bridges, A. C., J. Cocke, J. K. Olson, and R. T. Mayer: Effects of a new fluorescent insect growth regulator on the larval instars of *Aedes aegypti.* Mosq. News **37**, 227 (1977).

Broome, J. R., M. F. Callaham, L. A. Lewis, C. M. Ladner, and J. R. Heitz: The effects of rose bengal on the imported fire ant, *Solenopsis richteri* (Forel). Comp. Biochem. Physiol. **51C**, 117 (1975).

______ ______, W. E. Poe, and J. R. Heitz: Biochemical changes in the boll weevil induced by rose bengal in the absence of light. Chem. Biol. Interact. **14**, 203 (1976).

Butterworth, K. R., I. F. Gaunt, P. Grasso, and S. D. Gangolli: Acute and short-term toxicity studies on erythrosine BS in rodents. Food Cosmet. Toxicol. **14**, 525 (1976 a).

______ ______ ______ ______ Short-term toxicity of erythrosine BS in pigs. Food Cosmet. Toxicol. **14**, 533 (1976 b).

Callaham, M. F., L. A. Lewis, M. E. Holloman, J. R. Broome, and J. R. Heitz: Inhibition of the acetylcholinesterase from the imported fire ant, *Solenopsis richteri* (Forel), by dye-sensitized photooxidation. Comp. Biochem. Physiol. **51C**, 123 (1975 a).

______, J. R. Broome, O. H. Lindig, and J. R. Heitz: Dye-sensitized photooxidation reactions in the boll weevil, *Anthonomous grandis*. Environ. Entomol. **4**, 837 (1975 b).

______ ______, W. E. Poe, and J. R. Heitz: Time dependence of light-independent biochemical changes in the boll weevil, *Anthonomous grandis*, caused by dietary rose bengal. Environ. Entomol. **6**, 669 (1977 a).

______, C. O. Palmertree, J. R. Broome, and J. R. Heitz: Dye-sensitized photoinactivation of the lactic dehydrogenase and acetylcholinesterase from the boll weevil, *Anthonomous grandis*. Pestic. Biochem. Physiol. **7**, 21 (1977 b).

Carpenter, T. L., and J. R. Heitz: Light-dependent latent toxicity of rose bengal to *Culex pipiens quinquefasciatus*. Environ. Entomol. **9**, 533 (1980).

______, T. G. Mundie, J. H. Ross, and J. R. Heitz: Synergistic effect of fluorescein on rose bengal-induced, light-dependent toxicity. Environ. Entomol. **10**, 953 (1981 a).

______, and J. R. Heitz: Light-dependent and -independent toxicity of erythrosin B to *Culex pipiens quinquefasciatus* Say. Environ. Entomol. **10**, 972 (1981 b).

Certified Color Industry Committee: Use of certified FD&C colors in food. Food Technol. **22** (8), Table 3, p. 15 (1968).

Chung, K-T., G. E. Fulk, and A. W. Andrews: Mutagenicity testing of some commonly used dyes. Applied Environ. Microbiol. **42**, 641 (1981).

Clement, S. L., R. S. Schmidt, G. Szatmari-Goodman, and E. Levine: Activity of xanthene dyes against black cutworm larvae. J. Econ. Entomol. **73**, 390 (1980).

Collins, T. F. X., and E. L. Long: Effects of chronic oral administration of erythrosine in the mongolian gerbil. Food Cosmet. Toxicol. **14**, 233 (1976).

Colour Index: Publ. by The Society of Dyers and Colourists, Bradford, England and The American Association of Textile Chemists and Colorists. 3rd Ed. Lowell, MA (1971).

Conners, C. K.: Food additives and hyperactive children, p. 93. New York: Plenum Press (1980).

Creighton, C. S., T. L. McFadden, and J. M. Schalk: Toxicity of dietary rose bengal to larvae of the cabbage looper, corn earworm and pickleworm. J. Georgia Entomol. Soc. **15**, 66 (1980).

Daniel, J. W.: The excretion and metabolism of edible food colors. Toxicol. Applied Pharmacol. **4**, 572 (1962).

Daum, R. J., R. T. Gast, and T. B. Davich: Marking adult boll weevils with dyes fed in a cottonseed oil bait. J. Econ. Entomol. **62**, 943 (1969).

Dickinson, D., and T. W. Raven: Stability of erythrosine in artificially coloured canned cherries. J. Sci. Food Agr. **13**, 650 (1962).

Fairbrother, T. E.: Effects of xanthene dyes on face fly larvae and effects of erythrosin B on ruminant digestion. Ph.D. Thesis. Mississippi State University (1978).

______, H. W. Essig, R. L. Combs, and J. R. Heitz: Toxic effects of rose bengal and erythrosin B on three life stages of the face fly, *Musca autumnalis*. Environ. Entomol. **10**, 506 (1981).

Feingold, B. F.: Why your child is hyperactive. New York: Random House (1975).

Fondren, J. E., B. R. Norment, and J. R. Heitz: Dye-sensitized photooxidation in the house fly, *Musca domestica*. Environ. Entomol. **7**, 205 (1978 a).

______, and J. R. Heitz: Xanthene dye induced toxicity in the adult face fly, *Musca autumnalis*. Environ. Entomol. **7**, 843 (1978 b).

______ ______ Light intensity as a critical parameter in the dye-sensitized photooxidation of the house fly, *Musca domestica*. Environ. Entomol. **7**, 891 (1978 c).

______ ______ Dye-sensitized house fly toxicity produced as a function of variable light sources. Environ. Entomol. **8**, 432 (1979).

Foote, C. S.: Photosensitized oxidation and singlet oxygen: Consequences in biological systems. In W. A. Pryor (ed.): Free radicals in biology. Vol. 2, p. 85. New York: Academic Press (1976).

Goldberg, A. L., and R. J. Calvey: Automated high-performance liquid chromatographic determination of intermediates and side reaction products in FD&C Red No. 3. J. Assoc. Official Anal. Chemists **65**, 103 (1982).

Goldenring, J. R., D. K. Batter, and B. A. Shaywitz: Effect of chronic erythrosin B administration on developing rats. Neurobehav. Toxicol. Teratol. **3**, 57 (1981).

Graham, K.: Concepts of forest entomology, p. 256. New York: Reinhold (1963).

______ Entomological, ecological and evolutionary implications of photodynamic action. Can. J. Zool. **50**, 1631 (1972 a).

______, E. Wrangler, and L. H. Aasen: Susceptibility of the mealworm (*Tenebrio molitor* (L.)) to photodynamic injury by methylene blue. Can. J. Zool. **50**, 1625 (1972 b).

Gurr, E.: Synthetic dyes in biology, medicine and chemistry. New York: Academic Press (1971).

Gutter, B., W. T. Speck, and H. S. Rosenkranz: A study of the photoinduced mutagenicity of methylene blue. Mutat. Res. **44**, 177 (1977).

Haveland-Smith, R. B., and R. D. Combes: Screening of food dyes for genotoxic activity. Food Cosmet. Toxicol. **18**, 215 (1980).

______ ______, and B. A. Bridges: Studies on the genotoxicity of some fluorescein dyes. Mutat. Res. **88**, 1 (1981).

Hayes, D. K., and M. S. Schechter: Survival of codling moth larvae treated with methylene blue under short- and long-day photoperiods. J. Econ. Entomol. **63**, 997 (1970).

Heitz, J. R.: Erythrosin B kills flies in chicken manure. Interview; Chem. Eng. News **57** (17), 21 (1979).

______, and W. W. Wilson: Photodegradation of halogenated xanthene dyes. In M. V. Kennedy (ed.): Disposal and decontamination of pesticides, p. 35. Amer. Chem. Soc. Symposium series No. 73. Washington, D. C.: American Chemical Society (1978).

Hendricks, D. E.: Oil-soluble blue dye in larval diet marks adults, eggs and first-stage F_1 larvae of the pink bollworm. J. Econ. Entomol. **64**, 1404 (1971).

IARC; International Agency for Research on Cancer: Monographs on the evaluation of the carcinogenic risk of chemicals to man. Vol. 16, pp. 221,233. Lyon: IARC-WHO (1978).

Jodlbauer, A., and H. von Tappeiner: Ueber die Beteiligung des Sauerstoffes bei der photodynamischen Wirkung fluoreszierender Stoffe. Muench. Med. Wochenschr. **26**, 1139 (1904).

______ ______ Die Beteiligung des Sauerstoffes bei der Wirkung fluorescierender Stoffe. Dtsch. Arch. Klin. Med. **82**, 520 (1905).

Jones, R. L., E. A. Harrell, and J. W. Snow: Three dyes as markers for corn earworm moths. J. Econ. Entomol. **65**, 123 (1972).

Jordan, T. W., and J. N. Smith: Inhibition of house fly oxidative detoxication by phthaleins, fluoresceins and related compounds. Xenobiotica **11**, 1 (1981).

Kasha, M., and D. E. Brabham: Singlet oxygen electronic structure and photosensitization. Chapter 1 in Wasserman, H. H., and R. W. Murray (1979).

Kepka, A. G., and L. I. Grossweiner: Photodynamic inactivation of lysozyme by eosin. Photochem. Photobiol. **18**, 49 (1973).

Kon-Heitz: Exchange of letters between Kon, S. H. and J. R. Heitz. Chem. Eng. News **57** (27) 2; (31) 28; (45) 4 (1979).

Lavialle, M., and B. Dumortier: Effet photodynamique du bleu de méthylène sur les larves de *Pieris brassicae* (L.). C. R. Hebd. Seances Acad. Sci., Ser. D. **287**, 875 (1978).

Levitan, H.: Food, drug and cosmetic dyes: Biological effects related to lipid solubility. Proc. Nat. Acad. Sci. U.S.A. **74**, 2914 (1977).

Lewis, I. L., R. M. Patterson, and H. C. McBay: The effects of rhodamine B on the chromosomes of *Muntiacus muntjac*. Mutat. Res. **88**, 211 (1981).

Lillie, R. D.: H. J. Conn's Biological Stains. Baltimore: Williams & Wilkins (1969).

Lochmann, E-R., and A. Micheler: Molecular and biochemical aspects of photodynamic action. Photochem. Photobiol. **29**, 1199 (1979).

Logani, M. K., W. A. Austin, B. Shah, and R. E. Davies: Photooxidation of 8-methoxypsoralen with singlet oxygen. Photochem. Photobiol. **35**, 569 (1982).

Lutty, G. A.: The acute intravenous toxicity of biological stains, dyes and other fluorescent substances. Toxicol. Applied Pharmacol. **44**, 225 (1978).

Marccaci, A.: Sur l'action des alcaloides dans le règne végétale et animal. Arch.

Ital. Biol. **9**, 2 (1888).

Marmion, D. M.: Handbook of U.S. colorants for foods, drugs and cosmetics. New York: Wiley (1979).

Matoltsy, G., and Fábián, Gy.: Measurement of the photodynamic effect of cancerogenic substances with biological indicators. Nature **158**, 877 (1946).

Maugh, T. H.: Exploring plant resistance to insects. Science **216**, 722 (1982).

Maus, K. L., E. R. Nestmann, and D. J. Kowbel: Absence of mutagenicity of phloxine and phloxine B in *Escherichia coli* and in *Salmonella typhimurium*. Mutat. Res. **91**, 315 (1981).

Merkel, P. B., and D. R. Kearns: Radiationless decay of singlet molecular oxygen in solution. J. Amer. Chem. Soc. **94**, 7244 (1972).

McCann, J., and B. N. Ames: Detection of carcinogens as mutagens in the *Salmonella*/microsome test: Assay of 300 chemicals: Discussion. Proc. Nat. Acad. Sci. U.S.A. **73**, 950 (1976).

Nilsson, R., P. B. Merkel, and D. R. Kearns: Unambiguous evidence for the participation of singlet oxygen ($^1\Delta$) in photodynamic oxidation of amino acids. Photochem. Photobiol. **16**, 117 (1972).

Noonan, J. E., and H. Meggos: Synthetic food colors. In T. E. Furia (ed.): Handbook of food additives. 2nd ed., vol. 2, p. 339, Table 1. Boca Raton, FL: CRC Press (1980).

Peeples, W. A., and J. R. Heitz: The purification of xanthene dyes by reverse phase high performance liquid chromatography. J. Liq. Chromatog. **4**, 51 (1981).

Phillips, P. S., and C. A. Phillips: Photodynamic studies on acid ribonuclease from pea cotyledons. Experientia **37**, 1270 (1981).

Pimprikar, G. D., B. R. Norment, and J. R. Heitz: Toxicity of rose bengal to various instars of *Culex pipiens quinquefasciatus* and *Aedes triseriatus*. Environ. Entomol. **8**, 856 (1979).

———, J. E. Fondren, and J. R. Heitz: Small- and large-scale field tests of erythrosin B for house fly control in caged layer chicken houses. Environ. Entomol. **9**, 53 (1980 a).

———, B. L. Noe, B. R. Norment, and J. R. Heitz: Ovicidal, larvicidal and biotic effects of xanthene derivatives in the house fly, *Musca domestica* L. Environ. Entomol. **9**, 785 (1980 b).

Pooler, J. P., and D. P. Valenzeno: The role of singlet oxygen in photooxidation of excitable cell membranes. Photochem. Photobiol. **30**, 581 (1979).

Price, P. J., W. A. Suk, A. E. Freeman, W. T. Lane, R. L. Peters, M. L. Vernon, and R. J. Huebner: *In vitro* and *in vivo* indications of the carcinogenicity and toxicity of food dyes. Int. J. Cancer **21**, 361 (1978).

Raab, O.: Ueber die Wirkung fluorescirender Stoffe auf Infusorien. Z. für Biol. **39**, 524 (1900).

Respicio, N. C., and J. R. Heitz: Comparative toxicity of rhodamine B and rhodamine 6G to the house fly (*Musca domestica* L.). Bull. Environ. Contam. Toxicol. **27**, 274 (1981).

Rossi, E., A. van de Vorst, and G. Jori: Competition between singlet oxygen and electron transfer mechanisms in the porphyrin-sensitized photooxidation of L-tryptophan and tryptamine in aqueous cellular dispersions. Photochem. Photobiol. **34**, 447 (1981).

Sako, F., N. Kobayashi, H. Watabe, and N. Taniguchi: Cytotoxicity of food dyes on cultured fetal rat hepatocytes. Toxicol. Applied Pharmacol. **54**, 285 (1980).

Sakurai, H., and J. R. Heitz: Growth inhibition and photooxidative toxicity in the house fly, *Musca domestica* L., caused by xanthene dye in larval growth medium and after injection. Environ. Entomol. **11**, 467 (1982).

Schildmacher, H.: Über Photosensibilisierung von Stechmückenlarven durch fluoreszierende Farbstoffe. Biol. Zentralbl. **69**, 468 (1950).

Singh, A., and A. Petkau (eds.): Singlet oxygen and related species in chemistry and biology. Proc. Internat. Conf. on Singlet Oxygen, Pinawa, Manitoba, Canada, Aug. 1977. Photochem. Photobiol. **28**, 429 (1978).

Spikes, J. D.: Photodynamic action. In A. C. Giese (ed.): Photophysiology, vol. 3, p. 33. New York: Academic Press (1968).

______ Photosensitization. In K. C. Smith (ed.): The science of photobiology, p. 87. New York: Plenum Press (1977).

______, and B. W. Glad: Photodynamic action. Photochem. Photobiol. **3**, 471 (1964).

______, and R. Livingstone: The molecular biology of photodynamic action: sensitized photooxidations in biological systems. In L. G. Augenstein, R. Mason, and M. Zelle (eds.): Advances in radiation biology, vol. 3, p. 29. New York: Academic Press (1969).

Stern, A. M., V. D'Aurora, and D. S. Sigman: Inhibition of *Escherichia coli* DNA polymerase I by rose bengal. Arch. Biochem. Biophys. **202**, 525 (1980).

Sugimura, T., S. Sato, M. Nagao, T. Yahagi, T. Matsushima, Y. Seino, M. Takeuchi, and T. Kawachi: Overlapping of carcinogens and mutagens. In P. N. Magee, S. Takayama, T. Sugimura and T. Matsushima (eds.): Fundamentals in cancer prevention, p. 191. Baltimore: Univ. Park Press (1976).

Swanson, J. M., and M. Kinsbourne: Food dyes impair performance of hyperactive children on a laboratory learning test. Science **207**, 1485 (1980 a).

______, and W. J. Logan: Effects of food dyes on neurotransmitter accumulation in rat brain homogenate and on the behavior of hyperactive children. Effects of foods, drugs, dev. funct. nerv. syst., Methods Predict. Toxicity U.S. FDA., Publ. No. FDA-80-1076, p. 182 (1980 b).

Tappeiner, H. von, and A. Jodlbauer: Über die Wirkung der photodynamischen (fluorescierenden) Stoffe auf Protozoen und Enzyme. Dtsch. Arch. Klin. Med. **80**, 427 (1904).

Tonogai, Y., M. Iwaida, M. Tati, Y. Ose, and T. Sato: Biochemical decomposition of coal-tar dyes. Acute toxicity of coal-tar dyes and their decomposed products. J. Toxicol. Sci. **3**, 205 (1978).

______, Y. Ito, M. Iwaida, M. Tati, Y. Ose, and T. Sato: Studies on the toxicity of coal-tar dyes I. Photodecomposed products of four xanthene dyes and their acute toxicity to fish. J. Toxicol. Sci. **4**, 115 (1979 a).

______ ______ ______ ______ ______ ______ Studies on the toxicity of coal-tar dyes II. Examination of the biological reaction of coal-tar dyes to vital body. J. Toxicol. Sci. **4**, 211 (1979 b).

U.S. National Institutes of Health: Defined diets and childhood hyperactivity. Consensus development conference summary, vol. 4, No. 3.Bethesda, MD: NIH (1982).

Valenzeno, D. P., and J. P. Pooler: Phototoxicity: The neglected factor. J. Amer. Med. Assoc. **242**, 453 (1979).

______ ______ Cell membrane photomodification: Relative effectiveness of halogenated fluoresceins for photohemolysis. Photochem. Photobiol. **35**, 343 (1982).

Wade, M. J., and J. D. Spikes: The efficiency of halogenated fluoresceins as sensitizers for the photodynamic inactivation of trypsin. Photochem. Photobiol. **14**, 221 (1971).

Wasserman, H. H., and R. W. Murray (eds.): Singlet oxygen. New York: Academic Press (1979).

Weaver, J. E., L. Butler, and T. P. Yoho: Photodynamic action in insects: Volumetric changes in the hemolymph and crop contents of dye-treated, light-exposed cockroaches. Environ. Entomol. **5**, 840 (1976).

______ ______, and J. W. Amrine: Effects of erythrosin B on hemocytes of the American cockroach. Environ. Entomol. **11**, 463 (1982).

Wess, J. A., and D. L. Archer: Disparate *in vivo* and *in vitro* immunomodulatory activities of rhodamine B. Food Chem. Toxicol. **20**, 9 (1982).

Yoho, T. P., L. Butler, and J. E. Weaver: Photodynamic effect of light on dye-fed house flies: Preliminary observations of mortality. J. Econ. Entomol. **64**, 972 (1971).

______, J. E. Weaver, and L. Butler: Photodynamic action in insects. 1. Levels of mortality in dye-fed light-exposed house flies. Environ. Entomol. **2**, 1092 (1973).

______, L. Butler, and J. E. Weaver: Photodynamic killing of house flies fed food, drug and cosmetic dye additives. Environ. Entomol. **5**, 203 (1976).

Manuscript received October 12, 1982; accepted October 25, 1982.

Author's note added in proof:

Six reports of significant relevance to this review which have come to my attention after submission of the manuscript are:

Ballard, J. B., A. D. Vance, and R. E. Gold: Light-dependent and -independent response of German cockroach male and brown-banded nymph populations to two photodynamic dyes. Environ. Entomol., *in press* (1983).

Broome, J. R., M. F. Callaham, and J. R. Heitz: Xanthene dye-sensitized photooxidation in the black imported fire ant, *Solenopsis richteri*. Environ. Entomol. **4**, 883 (1975 b).

David, R. M., and J. R. Heitz: Toxicity of an imported fire ant bait based on phloxin B (D&C Red 27). J. Agr. Food Chem. **26**, 99 (1978).

Levitan, H.: Current issues in neurotoxicity; Food colors. Effects of foods, drugs, dev. funct. nerv. syst., Methods Predict. Toxicity. U.S. FDA., Publ. No. FDA-80-1076, p. 185 (1980).

Parkinson, T. M., and J. P. Brown: Metabolic fate of food colorants. Ann. Rev. Nut. **1**, 175 (1981).

Yoho, T.P.: The photodynamic effect of light on dye-fed house flies, *Musca domestica* L. Ph.D. Thesis. Univ. W. VA (1972).

Methyl bromide

By

George V. Alexeeff* and Wendell W. Kilgore*

Contents

*Department of Environmental Toxicology, University of California, Davis, CA 95616. The authors thank Jean Ackerman for her assistance in preparation of this document and translations.

Residue Reviews, Volume 88.

I. Introduction

The first case of human poisoning by methyl bromide was reported by Schuler in 1899. Since then over 950 poisoning cases involving fatalities, systemic poisonings, skin injuries, and eye injuries have been reported in the United States, Canada, England, France, Germany, Belgium, The Netherlands, Spain, Switzerland, India, Iraq, Japan, Australia, and the Soviet Union. The medical personnel reporting these incidents were particularly concerned about the potential danger of methyl bromide because of the absence of warning properties, the delayed onset of toxicity, the severity of symptoms, the poor prognosis of patients, and the variety of central nervous system (CNS) effects observed. Damage to other organs, particularly the respiratory system, has also been noted. These case histories comprise the bulk of toxicologic information available on methyl bromide. Consequently, little is known concerning methyl bromide's mechanism of toxic action in target organs or its actions on a cellular level. This information is not available although methyl bromide has been used extensively in agriculture for 50 years.

II. Uses

a) Early uses and applications

One of the first references to a suitable use for methyl bromide was made by Richardson (1897) when he proposed its use as a chemical agent in medicine to destroy malignant tissue. He advised that methyl bromide not be used as an anaesthetic, although effective as such, because it causes throat irritation and vomiting and easily liberates "free bromine." Apparently his caution was unheeded; in 1891 Richardson condemned the existing practice of using methyl bromide as an anaesthetic in dentistry, to which he attributed several deaths.

Methyl bromide has been economically useful since the early 1900s. The compound is a direct alkylating agent and reacts via a biomolecular nucleophilic substitution. It has been used as a methylating agent in the chemical industry, especially in the preparation of aniline dyes and antipyrine (Henning 1933,

Mackie 1938, Irish *et al.* 1940, Miller 1943, Clarke *et al.* 1945, Wyers 1945, von Oettingen 1946, Stenger 1978). Current use of methyl bromide in this area is limited; methyl iodide is usually preferred for its higher boiling point.

The boiling point of methyl bromide is 3.6°C at 760 torr (Weast 1981-1982), which suggests it might be suitable for use as a refrigerant. Patents were obtained for such use, either alone or mixed with methyl chloride (Chandler 1925, Jones 1928). Sayers *et al.* (1929) and Nuckolls (1933) conducted investigations of the general toxic effects of methyl bromide to determine its suitability as a refrigerant. Probably as a result of these reports of its potential toxicity, the preferred use of methyl chloride in refrigeration and the development of other refrigerants, methyl bromide never became a common refrigerant in the United States.

Methyl bromide is nonflammable, is roughly 3.5 times heavier than air, and vaporizes quickly. These properties were utilized when methyl bromide was promoted as a fire extinguishing agent. The compound was recommended for general use as a fire extinguisher in England by Henning (1933) because it was so effective and did not produce phosgene as did carbon tetrachloride, the other fire extinguisher in general use at that time. Wyers summarized its usefulness in 1945:

> The principle use to which methyl bromide is put in [England] is that of extinguishing fires. Bulk for bulk, less is required for that purpose than any other known substance. It has, further, the advantage of being self-trajectory. A mixture with carbon tetrachloride is frequently used. Especially is methyl bromide recommended for fires involving celluloid or volatile hydrocarbons.

Methyl bromide had considerable use as a fire extinguishing agent in the 1920s through the 1940s, particularly in Europe (Wyers 1945, Butler *et al.* 1945, Clarke *et al.* 1945, Auck 1972, Botteri *et al.* 1972, Ford 1975). The methyl bromide fire extinguisher was introduced into the United States as the "standard of Europe" in the late 1920s. It was marketed in the United States in small fire extinguishers for a period of time, but it was never accepted as a legitimate fire extinguishing agent due to a toxicologic study conducted by Underwriters' Laboratories, Inc. in 1932. Instead, carbon tetrachloride remained the standard agent in the United States until it was replaced by chlorobromomethane and bromotrifluoromethane in the 1950s and 1960s (Auck 1972). In England the reports of methyl bromide's toxicity were questioned and impurities were implicated as the actual toxicants (Henning 1933). Methyl bromide became the standard agent for marine and aircraft fire protection systems in England and Germany in the late 1930s. It is thought still to be in use today in some of the British aircraft air extinguishing systems (Ford 1975).

b) Present uses—fumigation

The insecticidal properties of methyl bromide were first noticed by Le Goupil (1932) when he added methyl bromide as a fire retardant to the fumigant ethylene oxide. The increase in this mixture's effectiveness as an insecticide led him to perform tests on the insecticidal properties of methyl bromide alone. However, pure methyl bromide could not be used as a fumigant until new cylinders were designed (Mackie 1938). During the first few years it was used as a fumigant in the United States, it was mixed with carbon dioxide. It is interesting to note that increasing concern over methyl bromide's toxicity has led to efforts to reduce exposure to it. In 1974 Snitko *et al.* reported the advantages of a new fumigant which was a mixture of methyl bromide and carbon dioxide. Although it was not a new discovery, such a mixture may even now be suitable for many fumigant applications and could reduce the number of methyl bromide poisoning cases.

Since methyl bromide is colorless and has little odor at potentially toxic concentrations, severe exposures can occur unknowingly. Sayers *et al.* (1929) suggested that a warning agent should be added to methyl bromide refrigerants to signify the presence of a leak. In 1946 von Oettingen noted that although suggestions had been made to add an odorant to methyl bromide fire extinguishers, such a signal might not be detected in time due to methyl bromide's inherent volatility. In 1949 Dieter and Coulter reported that 2% chloropicrin could be added as a warning agent without impairing the effectiveness of methyl bromide as a soil fumigant. They cautioned that although chloropicrin may be useful in cases of accidental spillage or leakage, after a few hours its odor may no longer be detectable. Van Assche (1971) questioned the actual safety of using a mixture with only 2% chloropicrin. It should be noted that methyl bromide's vapor pressure is 75 times that of chloropicrin at 20°C (1,380 mmHg and 18.3 mmHg, respectively). Van Assche indicated that chloropicrin irritation is sensed at 1.3 ppm and that a level of 0.44 mg/L (115 ppm) of methyl bromide may be attained under these circumstances. Formulations with more chloropicrin are available, such as Dowfume MC-33 with 33% chloropicrin. The California Department of Food and Agriculture (1974) recommended in a pesticide safety bulletin that methyl bromide should be used in formulations with chloropicrin whenever possible.

The most extensive reviews on the applications of methyl bromide as a fumigant are those by Thompson (1966), Harry and Brown (1974), and White (1978). In brief, methyl bromide has been used for a variety of pests including insects (Mackie and Carter 1937, Mackie 1938, Fisk 1938, Dudley and Neal 1942, Bond 1978), nematodes (Hague and Sood 1963), rodents (Berry 1938), bacteria and viruses (Schmittle 1955, Maag and Schmittle 1962), fungi (Munnecke *et al.* 1978, Menge *et al.* 1978), mites (Powell 1977, Bowley 1981), and weeds (Skimina 1977, White 1978). It has been found useful for fumigation of stored food commodities in barges, in ships' holds, in fumigation chambers, in grain silos, in flour mills, in warehouses, and under "gas-proof" sheets (Thompson 1966). Another major use is "sterilization" of soil covered with "gas-proof" sheeting (Hague and Sood 1963), particularly prior to the planting of strawberries, tomatoes, tobacco, and Christmas trees (Van Gundy and Van Gundy 1971). Similarly, in the mushroom industry it is used to disinfect compost (Harry and Brown 1974). Other uses include fumigation of buildings (Mazel *et al.* 1946, Baynes-Cope 1972), furniture (Thompson 1966), and books and archives (Baynes-Cope 1972, Gallo 1978 a and b).

The annual production of methyl bromide in the United States, according to the latest information available (USDA 1970 and 1978), is about 35 million lb (Table I). The average amount of methyl bromide used annually in California in the last 11 yr is approximately 7.9 million lb (Table I), which constitutes about 22.5% of the annual U.S. production. This makes methyl bromide one of the most widely used organic pesticides in California. In fact, in 1980 the only pesticides used in greater quantity than methyl bromide were sulfur (26.8 million lb), D-D mixture (13.4 million lb), and petroleum oil (11.7 million lb). The increase in the volume of use and in the variety of uses of methyl bromide as a fumigant is in large part attributable to its high toxicity to many pests, the variety of settings in which it can be applied, its ability to penetrate the fumigated substances, and its rapid dissipation following application.

Table I. *Annual production of methyl bromide in the United States and annual consumption in California.*

Year	Produced (million lb)[a]	Used in California (million lb)[b]
1960	12.659	NA
1961	12.892	NA
1962	12.757	NA
1963	17.394	NA
1964	16.994	NA
1965	14.303	NA
1966	16.345	NA
1967	19.454	NA
1968	20.454	NA
1969	20.033	NA
1970	21.047	3.102
1971	NA	3.237
1972	24.633	5.368
1973	29.571	36.649[c]
1974	30.452	0.521[c]
1975	36.048	7.164
1976	35.856	4.182
1977	34.684	6.553
1978	NA	5.538
1979	NA	8.371
1980	NA	6.064
1981	NA	6.257

[a] United States Department of Agriculture, 1970 and 1978.
[b] California Department of Food and Agriculture, 1970 to 1981.
[c] Data for this year are under question.

c) Uses associated with human poisoning incidents

The potential for exposure of human beings to methyl bromide is directly related to the manufacture and uses of the compound. Methyl bromide poisoning incidents reported through 1952 were enumerated by von Oettingen (1955). He reported that the greatest number of methyl bromide poisoning incidents resulted from chemical manufacture and filling operations, followed by fire extinguishers and fumigation (Table II). Since 1955 fumigation has become the major source of fatalities and systemic poisonings due to methyl bromide.

A number of poisoning incidents have occurred in the manufacture, filling, storage, transportation, and disposal of methyl bromide (Schuler 1899, Jaquet 1901, Cade and Mazel 1923, Duvoir *et al.* 1937, Mackie 1938, Watrous 1942, Davay 1972, *San Francisco Chronicle* 1973, Maddy *et al.* 1981). These reports include 17 fatalities, 153 systemic illnesses, and 30 skin injuries (Table II).

The first report describing the potential dangers of methyl bromide as a fire extinguisher was made in 1930 by Joachimoglu. Even so, the use of methyl bromide as a fire extinguisher produced a number of reported poisoning cases due to accidental discharge, leakage, or prolonged exposure during use (Tietze 1933, Duvoir *et al.* 1937, De Gaulejac and Dervillee 1939, Billet and Abel 1940, Derobert 1941, Roger and Hawthorn 1941, Florentin 1944, Gray 1944, Holling and Clarke 1944, Clarke *et al.* 1945, Viner 1945, Butler *et al.* 1945, Wyers 1945, Carter 1945, Benatt and Courtney 1948, Gayral 1949, MacDonald *et al.* 1950, Irsigler 1951, Marchand-Aliphant 1952, Prain and Smith 1952, Derobert *et al.* 1953, Beumie 1953, Hamon *et al.* 1955, Sigwald *et al.* 1955, Roche *et al.* 1958, Andrieu *et al.* 1959, Pernod *et al.* 1961, Grossetette *et al.* 1963, Longley and Jones 1965, Castaing *et al.* 1966, Mellerio *et al.* 1974). These reports and those summarized by von Oettingen (1955) indicate there have been at least 34 deaths, 64 systemic poisonings, and 10 skin injuries associated with the use of methyl bromide as a fire extinguisher (Table II).

The use of methyl bromide as a refrigerant has not been particularly related to toxic exposure. Hunter (1975) alluded to two deaths in an apartment house caused by methyl bromide vapor escaping from a refrigerator and indicated that other poisoning cases have occurred in the process of charging and repairing refrigerators.

By far the largest number of reported methyl bromide poisoning cases have occurred during the fumigation process, injuring either the fumigator or a bystander (Mackie 1938, Schifferli 1942, Miller 1943, Johnstone 1945, Mazel *et al.* 1946, Ingram 1951, Jordi 1953, Corcos *et al.* 1955, Miller *et al.* 1961, Rathus and Landy 1961, Kantarjian and Shaheen 1963, Drawneek *et al.* 1964, Collins 1965, Hine 1969, Greenberg 1971, Takahashi *et al.* 1973, U.S. HEW 1974, Garcia Rico *et al.* 1974, Mellerio *et al.* 1974, Maddy 1976, Toyanaga and Tokuda 1976, Bezuglyi *et al.* 1977, Shield *et al.* 1977, Ushio and Osozuka 1977, Yomura and Shibata 1978, Ishizu *et al.* 1979, Jameson 1979, Takayama and Numajiri 1979, Maddy *et al.* 1981, Zatuchni and Hong 1981). Fumigation has been associated

Table II. *Methyl bromide poisoning incidents.*

A. Resulting from chemical manufacture, filling, storage, transportation, or disposal.

Severity	*No. of incidents*	*Source*
Fatalities	16	von Oettingen (1955)
Fatality	1	Davay (1972)
Total	17	
Systemic poisoning	108	von Oettingen (1955)
Systemic poisoning	1	Longley & Jones (1965)
Systemic poisoning	1	Davay (1972)
Systemic poisoning	18	S. F. Chronicle (1973)
Systemic poisoning	25	Maddy *et al.* (1981)
Total	153	
Skin injuries only	26	von Oettingen (1955)
Skin injuries only	4	Maddy *et al.* (1981)
Total	30	

B. Resulting from use or leaking fire extinguishers.

Severity	*No. of incidents*	*Source*
Fatality	1	Marchand-Aliphant (1952)
Fatalities	4	Beumie (1953)
Fatality	1	Derobert *et al.* (1953)
Fatalities	18	von Oettingen (1955)
Fatality	1	Roche *et al.* (1958)
Fatality	1	Franken *et al.* (1959)
Fatality	1	Pernod *et al.* (1961)
Fatality	1	Langlois *et al.* (1963)
Fatalities	3	Mosinger *et al.* (1964)
Fatalities	2	Castaing *et al.* (1966)
Fatalities	2	Mellerio *et al.* (1974)
Total	35	
Systemic poisoning	9	de Gaulejac & Dervillee (1939)
Systemic poisoning	2	Roger & Hawthorne (1941)
Systemic poisoning	1	Marchand-Aliphant (1952)
Systemic poisoning	2	Beumie (1953)
Systemic poisoning	1	Hamon *et al.* (1955)
Systemic poisoning	1	Sigwald *et al.* (1955)
Systemic poisoning	22	von Oettingen (1955)
Systemic poisoning	2	Andrieu *et al.* (1959)
Systemic poisoning	1	Bernot & Goldberg (1960)

Table II (*continued*)

Systemic poisoning	12	Pernod *et al.* (1961)
Systemic poisoning	5	Grossetette *et al.* (1963)
Systemic poisoning	3	Langlois *et al.* (1963)
Systemic poisoning	2	Castaing *et al.* (1966)
Systemic poisoning	3	Goulon *et al.* (1975)
Total	66	

Severity	*No. of incidents*	*Source*
Skin injuries only	4	de Gaulejac & Dervillee (1939)
Skin injuries only	6	von Oettingen (1955)
Total	10	

C. Resulting from uses as a fumigant.

Severity	*No. of incidents*	*Source*
Fatality	1	Jordi (1953)
Fatality	1	Corcos *et al.* (1955)
Fatalities	42	von Oettingen (1955)
Fatalities	5	Hine (1969)
Fatalities	7	Calif. Dept. of Health (1973)
Fatalities	2	Garcia Rico *et al.* (1974)
Fatalities	2	Ishizu *et al.* (1969)
Total	60	
Systemic poisoning	1	Jordi (1953)
Systemic poisoning	2	von Oettingen (1955)
Systemic poisoning	108	Calif. Dept. of Health (1957-73)
Systemic poisoning	7	Rathus & Landy (1961)
Systemic poisoning	8	Kantarjian & Shaheen (1963)
Systemic poisoning	1	Drawneek *et al.* (1964)
Systemic poisoning	1	Collins (1965)
Systemic poisoning	5	Hine (1969)
Systemic poisoning	1	Kashima *et al.* (1969)
Systemic poisoning	2	Williams & Reilly (1970)
Systemic poisoning	14	Araki *et al.* (1971)
Systemic poisoning	1	Eckardt (1971)
Systemic poisoning	1	Greenberg (1971)
Systemic poisoning	1	Takahashi *et al.* (1973)
Systemic poisoning	11	Garcia Rico *et al.* (1974)
Systemic poisoning	2	Mellerio *et al.* (1974)
Systemic poisoning	3	U.S. Dept. HEW (1974)
Systemic poisoning	14	Maddy (1976)
Systemic poisoning	1	Toyonaga & Tokuda (1976)
Systemic poisoning	2	Bezuglyi (1977)

Table II (*continued*)

Systemic poisoning	3	Shield *et al.* (1977)
Systemic poisoning	1	Ushio & Osozuka (1977)
Systemic poisoning	1	Yomura & Shibata (1978)
Systemic poisoning	1	Ishizu *et al.* (1979)
Systemic poisoning	4	Takayama & Numajiri (1979)
Systemic poisoning	104	Maddy *et al.* (1981)
Systemic poisoning	1	Zatuchni & Hong (1981)
Total	301	
Skin injuries	90	Calif. Dept. of Health (1957-73)
Skin injuries	10	Maddy (1976)
Skin injuries	45	Maddy *et al.* (1981)
Total	145	
Eye injuries	20	Calif. Dept. of Health (1957-73)
Eye injuries	4	Maddy (1976)
Eye injuries	33	Maddy *et al.* (1981)
Total	57	
Other or unspecified injuries	63	Calif. Dept. of Health (1952-73)
Other or unspecified injuries	15	Dubrisay & Fages (1978)
Total	78	

D. Unspecified cases.

Severity	*No. of incidents*	*Source*
Fatalities	3	Gaultier *et al.* (1973)
Systemic poisoning	1	Thiebaut *et al.* (1943)
Systemic poisoning	2	Gaultier *et al.* (1973)
Total	6	

Grand Totals

Fatalities	115
Systemic injuries	523
Skin injuries	185
Eye injuries	57
Other or unspecified	78

with at least 60 fatalities, 301 systemic poisonings, 145 skin injuries, 57 eye injuries, and 78 other unspecified injuries (Table II). Although a majority of the methyl bromide poisoning cases summarized occurred in California, in this state methyl bromide has produced fewer illnesses than parathion over the last 25 yr (California Department of Health 1957-1973, Maddy 1976, Maddy *et al.* 1981). In California, methyl bromide has reportedly produced 236 systemic illnesses, 144 skin irritation cases, 57 eye injuries, and 69 other or unspecified injuries. In contrast, over the same period parathion produced 1,521 systemic illnesses, 90 skin irritation cases, 42 eye injuries, and 246 other or unspecified cases. Thus, parathion has produced almost four times (1899) as many poisoning incidents as methyl bromide (506) in California.

III. Route of exposure

a) Animal

Most experimental data obtained from exposure of laboratory animals to methyl bromide were reviewed by von Oettingen (1946). In that publication he summarized the investigations that had been performed in dogs (Merzback 1928, Beyne and Goett 1934, Duvoir *et al.* 1939), cats (Glaser and Frisch 1929), rabbits (Roth and von Frey, quoted in Schuler 1899, Glaser and Frisch 1929, Beyne and Goett 1934, Duvoir *et al.* 1939), mice (Roth and von Frey, quoted in Schuler 1899, Bachem 1927, Glaser and Frisch 1929), rats (Winter 1936, Irish *et al.* 1940), and guinea pigs (Roth and von Frey, quoted in Schuler 1899, Sayers *et al.* 1929, Nuckolls 1933, Kohn-Abrest, quoted in Beyne and Goett 1934). Von Oettingen (1946) pointed out several difficulties in evaluating these studies including the great variability in experimental conditions, the frequent absence of methyl bromide analysis during exposure, and the small number of animals used in most experiments. He concluded that only the experiments in guinea pigs by Sayers *et al.* (1929) and in rats and rabbits by Irish *et al.* (1940) had sufficient number of animals to determine a "minimum fatal or maximal tolerated concentration." Laboratory animal studies that have recently appeared in the literature include those by Ikeda *et al.* (1980), Miyagawa (1982), Miyagawa *et al.* (1980), Homma *et al.* (1980), and Anger *et al.* (1981). Several of these studies considered the behavioral and neurological effects of the compound.

The major acute studies in laboratory animals were conducted in guinea pigs by Sayers *et al.* (1929) and in rats and rabbits by Irish *et al.* (1940). In these experiments, animals were exposed to a specific concentration of methyl bromide until observable death, or for a prescribed length of time and then removed for observations. A study by Miyagawa *et al.* (1980) exposed rats to various concentrations of methyl bromide for 24 hr. Subchronic and chronic studies have been reported by Irish *et al.* (1940), Ikeda *et al.* (1980), and Anger *et al.* (1981). Daily methyl bromide exposures for four hr a day, five days a wk, were administered to rats by Ikeda *et al.* (1980) for three wk. A one-mon exposure study in rats and rabbits was conducted by Anger *et al.* (1981). In their study the animals were exposed 7.5 hr/day, four days/wk, to achieve a total 100-hr exposure period. In a separate experiment, Anger *et al.* (1981) exposed rats to methyl bromide six hr/day, five days a wk for a period of 12 mon, to achieve a total 1,080-hr exposure period. In the study by Irish *et al.* (1940), rats, rabbits, guinea pigs, and monkeys were exposed 7.5 to eight hr/day, five days/wk for up to six mon.

The laboratory animal studies of Sayers *et al.* (1929), Irish *et al.* (1940), Ikeda *et al.* (1980), and Anger *et al.* (1981) were all performed by whole body inhalation. No dermal experiments with methyl bromide have been reported. A feeding study of methyl bromide in olive oil was performed with rabbits by Dudley and Neal (1942).

b) Human

There are two major routes of exposure for methyl bromide: dermal and inhalation. Dermal exposure to liquid methyl bromide resulting from accidental splashing or other contact has been reported in a variety of circumstances in the last 45 yr (Mackie 1938, De Gaulejac and Dervillee 1939, Billet and Abel 1940, Schifferli 1942, Watrous 1942, Gray 1944, Wyers 1945, Butler *et al.* 1945, Thomson 1945, Benatt and Courtney 1948, Jordi 1953, Corcos *et al.* 1953, Longley and Jones 1965, Maddy *et al.* 1981). In addition to direct contact, exposure can occur through contact with contaminated boots, clothing, bandages, or gloves. These articles are often constructed of materials such as rubber that can absorb methyl bromide (Watrous 1942, Gray 1944, Wyers 1945, Butler *et al.* 1945, Benatt and Courtney 1948, Jordi 1953, Maddy *et al.* 1981). Direct eye injury or irritation has also occurred (Holling and Clarke 1944, Wyers 1945, Clarke *et al.* 1945, Maddy *et al.* 1981).

The primary route of exposure to methyl bromide is inhalation. Poisonings have occurred following acute, subchronic, and chronic exposures. Acute cases involved spilling or leaking while handling methyl bromide and exposure of persons unaware of its presence (De Gaulejac and Dervillee 1939, Billet and Abel 1940, Roger and Hawthorn 1941, Watrous 1942, Schifferli 1942, Miller 1943, Holling and Clarke 1944, Gray 1944, Florentin 1944, Viner 1945, Butler *et al.* 1945, Clarke *et al.* 1945, Thomson 1945, Benatt and Courtney 1948, Mazel *et al.* 1946, MacDonald *et al.* 1950, Irsigler 1951, Marchand-Aliphant 1952, Prain and Smith 1952, Beumie 1953, Jordi 1953, Hamon *et al.* 1955, Sigwald *et al.* 1955, Andrieu *et al.* 1959, Pernod *et al.* 1961, Longley and Jones 1965, Castaing *et al.* 1966, Hine 1969, Davay 1972, *San Francisco Chronicle* 1973, Garcia Rico *et al.* 1974, Goulon *et al.* 1975, Bezuglyi *et al.* 1977, Shield *et al.* 1977, Takayama and Numajiri 1979). Acute cases include both well-documented single-exposure incidents and incidents where the potential for exposure were very short (less than two wk), as might have occurred with a new worker. Cases of exposure lasting two wk to three mon are assumed to be subchronic. Relatively few such cases have been documented (Dejong 1944, Carter 1945, Kantarjian and Shaheen 1963, Eckardt 1971). If the subchronic exposure ended with a large single exposure, then the toxic effects were similar to those of acute cases. Otherwise, the subchronic exposure syndromes resembled those of chronic toxicity and these two will be considered together. An exposure was considered chronic when the potential or likelihood of exposure was three mon or longer in duration (De Gaulejac and Dervillee 1939, Dejong 1944, Florentin 1944, Viner 1945, Wyers 1945, Carter 1945, Johnstone 1945, Gayral 1949, Ingram 1951, Corcos *et al.* 1955, Franken *et al.* 1959, Rathus and Landy 1961, Drawneek *et al.* 1964, Collins 1965, Hine 1969, Williams and Reilly 1970, Greenberg 1971, Araki *et al.* 1971, Yoshida *et al.* 1975, Ushio and Osozuka 1977, Yomura and Shibata 1978, Zatuchni and Hong 1981). In some cases symptoms became evident after continuous, generally unnoticed, chronic exposure. In other cases the exposure victim had a chronic history of handling methyl bromide with the potential for exposure, but the onset of symptoms only occurred following an unusually high exposure. Since methyl bromide does not accumulate in the body, the chronic effects are probably the result of an accumulation of the lesion.

IV. Dermal toxicity

Dermal contact with methyl bromide immediately causes a tingling and cool sensation, sometimes with numbness or aching pain (De Gaulejac and Dervillee 1939, Watrous 1942). The exposure incident may result in redness, congestion, dermatitis, itching, swollen areas, and blistering with gelatinous material underneath (De Gaulejac and Dervillee 1939, Billet and Abel 1940, Schifferli 1942, Watrous 1942, Dejong 1944, Gray 1944, Wyers 1945, Thomson 1945, Carter 1945, Butler *et al.* 1945, Benatt and Courtney 1948, Jordi 1953, Heise and Moller 1976, Maddy *et al.* 1981). Aside from skin irritation, dermal exposure to methyl bromide has been associated with systemic effects typical of inhalation of the compound (De Gaulejac and Dervillee 1939, Billet and Abel 1940, Schifferli 1942, Watrous 1942, Gray 1944, Wyers 1945, Thomson 1945, Benatt and Courtney 1948, Jordi 1953, Longley and Jones 1965, Maddy *et al.* 1981). These effects will be discussed with acute inhalation exposure.

V. Inhalation exposure concentration and mortality

a) Animal

The mortality of laboratory animals following acute exposure to methyl bromide is presented in Table III. The Sayers *et al.* (1929) study reported 100, 67, 17, and 0% mortality following a variety of exposure conditions. The Irish *et al.* (1940) study reported 100 and 0% mortality exposure conditions. The Miyagawa *et al.* (1980) study reported several mortality figures, from which a 24-hr LC_{50} in rats can be estimated to be 0.05 mg/L (13 ppm) of methyl bromide.

Table III. *Fatal and non-fatal inhalation concentrations following acute exposure of laboratory animals to methyl bromide.*

Mortality (%)	Species	N[a]	Concentration (mg/L)	Time (hr)[b]	Time of death	Source
100	g. pig	3	370.3	0.25	exp	Sayers *et al.* 1929
100	g. pig	2	192.9	0.25	exp	Sayers *et al.* 1929
100	g. pig	2	111.9	0.08	24h	Sayers *et al.* 1929
100	g. pig	4	84.9	0.58	exp	Sayers *et al.* 1929
100	g. pig	6	84.9	0.17	24h	Sayers *et al.* 1929
100	g. pig	2	65.6	0.17	24h	Sayers *et al.* 1929
100	g. pig	2	50.1	1.17	exp	Sayers *et al.* 1929
100	g. pig	4	27.0	0.50	24h	Sayers *et al.* 1929
100	g. pig	6	27.0	1.5	exp	Sayers *et al.* 1929
100	g. pig	2	20.8	0.33	6d	Sayers *et al.* 1929
100	g. pig	4	8.9	1.5	24h	Sayers *et al.* 1929
100	g. pig	6	8.9	2.83	exp	Sayers *et al.* 1929
100	g. pig	4	2.3	4.5	48h	Sayers *et al.* 1929
100	g. pig	6	1.9	8	exp	Sayers *et al.* 1929
100	g. pig	6	1.2	9	3d	Sayers *et al.* 1929
100	g. pig	6	1.2	13.5	3d	Sayers *et al.* 1929
100	rat	NA	50	0.10	NA	Irish *et al.* 1940
100	rat	NA	20	0.40	NA	Irish *et al.* 1940
100	rat	NA	10	0.70	NA	Irish *et al.* 1940
100	rat	NA	2.0	6	NA	Irish *et al.* 1940
100	rat	NA	1.0	22	NA	Irish *et al.* 1940
100	rat	NA	0.85	26	NA	Irish *et al.* 1940
100	rabbit	NA	50	0.5	NA	Irish *et al.* 1940
100	rabbit	NA	20	1.4	NA	Irish *et al.* 1940

Table III *(continued)*

Mortality (%)	Species	N[a]	Concentration (mg/L)	Time (hr)[b]	Time of death	Source
100	rabbit	NA	10	2.2	NA	Irish *et al.* 1940
100	rabbit	NA	2.0	11	NA	Irish *et al.* 1940
100	rabbit	NA	1.0	24	NA	Irish *et al.* 1940
100	rabbit	NA	0.85	32	NA	Irish *et al.* 1940
100	rat	5	0.46	24	exp	Miyagawa *et al.* 1980
100	rat	5	0.39	24	exp	Miyagawa *et al.* 1980
100	rat	5	0.31	24	24h	Miyagawa *et al.* 1980
100	rat	5	0.23	24	24h	Miyagawa *et al.* 1980
80	rat	5	0.15	24	48h	Miyagawa *et al.* 1980
80	rat	5	0.075	24	96h	Miyagawa *et al.* 1980
75	g. pig	4	0.58	9.0	3d	Sayers *et al.* 1929
50	g. pig	2	1.2	5.0	3d	Sayers *et al.* 1929
40	rat	5	0.039	24	8d	Miyagawa *et al.* 1980
25	g. pig	4	8.9	0.5	24h	Sayers *et al.* 1929
0	g. pig	2	7.7	0.5	–	Sayers *et al.* 1929
0	g. pig	2	2.3	1.5	–	Sayers *et al.* 1929
0	g. pig	4	1.2	4.5	–	Sayers *et al.* 1929
0	g. pig	2	0.58	13.5	–	Sayers *et al.* 1929
0	g. pig	4	0.39	5.0	–	Sayers *et al.* 1929
0	g. pig	3	0.39	10.0	–	Sayers *et al.* 1929

0	rat	NA	50	0.5	–	Irish *et al.* 1940
0	rat	NA	20	0.1	–	Irish *et al.* 1940
0	rat	NA	10	0.4	–	Irish *et al.* 1940
0	rat	NA	2.0	2.0	–	Irish *et al.* 1940
0	rat	NA	1.0	8.0	–	Irish *et al.* 1940
0	rat	NA	0.85	12	–	Irish *et al.* 1940
0	rat	NA	0.42	22	–	Irish *et al.* 1940
0	rabbit	NA	50	0.2	–	Irish *et al.* 1940
0	rabbit	NA	20	0.6	–	Irish *et al.* 1940
0	rabbit	NA	10	1.0	–	Irish *et al.* 1940
0	rabbit	NA	20	6.0	–	Irish *et al.* 1940
0	rabbit	NA	1.0	8.33	–	Irish *et al.* 1940
0	rabbit	NA	0.85	20	–	Irish *et al.* 1940

[a] Animals sacrificed before 7 days are not included.
[b] NA = not available, exp = death during exposure, h = hr, and d = days.

Bakhishev (1973) reported that the LC_{50} in rats following a 30-min exposure to methyl bromide is 11.0 mg/L. Dawson *et al.* (1977) determined that the 96-hr LC_{50} of methyl bromide was 4.18 mg/L in bluegill sunfish (*Lepomis macrochirus*), a fresh-water fish, and 4.68 mg/L in tidewater silversides (*Menidia beryllia*), a salt-water fish.

In the repeated exposure experiments of Ikeda *et al.* (1980) 3 of 12 animals died during exposure to 300 ppm (1.16 mg/L) of methyl bromide. In the one-mon experiments by Anger *et al.* (1981), no fatalities occurred following exposure of rabbits and rats to 65 ppm (0.25 mg/L). In their 36-wk study, rats were exposed to 55 ppm (0.21 mg/L); 6 of 32 exposed rats and 3 of 8 control rats died of undetermined causes. When Irish *et al.* (1940) repeatedly exposed animals to 0.85 mg/L of methyl bromide, the rats, guinea pigs, and rabbits died after a few exposures. At exposures of 0.42 mg/L, five of 30 rats survived 83 to 100 exposures, and seven of 11 guinea pigs survived to the end of the 6-mon experiment. When concentrations of 0.25 mg/L of methyl bromide were used, one of 22 rats and one of 24 guinea pigs succumbed, while 14 of 42 rabbits died (the others were sacrificed while showing severe symptoms). Exposure to 0.13 mg/L did not produce any fatalities in rats, guinea pigs, monkeys, or rabbits, but rabbits did exhibit signs of toxicity. Exposure of rabbits to a concentration of 0.065 mg/L did not produce any fatalities or significant observable effects.

For the mortality response, the results of continuous exposure appear to be the same as that of repeated exposures. Irish *et al.* (1940) observed 100% mortality following an exposure of rats and rabbits to 0.85 mg/L of methyl bromide. All rats succumbed following a 26-hr continuous exposure or 23 to 32 hr of repeated exposures. All rabbits died following a 32-hr continuous exposure or following very few repeated 7.5- to 8-hr exposures (probably a total of 24 to 32 hr). Thus, the 16-hr recovery period is not sufficient to prevent mortality following repeated exposure to this concentration.

According to available information, the rabbit is the most sensitive laboratory animal species to methyl bromide toxicity. The data further indicate that a no-observable-effect level in rabbits is roughly 0.065 mg/L (17 ppm) for an 8-hr daily exposure to methyl bromide for 6 mon (Irish *et al.* 1940).

b) Human

Fatal exposures to methyl bromide have been reported in the literature on numerous occasions (Table II). Death usually occurred from either acute exposures to very high concentrations (Duvoir *et al.* 1937, Miller 1943, Holling and Clarke 1944, Florentin 1944, Wyers 1945, Clarke *et al.* 1945, Viner 1945, Mazel *et al.* 1946, MacDonald *et al.* 1950, Prain and Smith 1952, California Department of Health 1958 to 1965, Pernod *et al.* 1961, Hine 1969, Davay 1972, Garcia Rico *et al.* 1974, Mellerio *et al.* 1974) or following higher than usual exposure to methyl bromide on top of a history of chronic low-level exposure (Schifferli 1942, Florentin 1944, Jordi 1953, Franken *et al.* 1959

California Department of Health 1959 and 1960, Hine 1969). For the acute incidents, some of the exposure estimates are 231.4 mg/L for two-hr (Wyers 1945), 30.9 mg/L for one-and-one-half hr (Schifferli 1942) to six-and-one-half hr (Miller 1943), and 6.2 mg/L for about 10 to 20 hr of exposure (Holling and Clarke 1944). Whatever the concentration, the deaths that resulted from human exposure to methyl bromide usually followed exposure periods of at least one-and-one-half hr and usually several hr (Duvoir *et al.* 1937, Schifferli 1942, Miller 1943, Florentin 1944, Wyers 1945, Viner 1945, Clarke *et al.* 1945, Mazel *et al.* 1946, Prain and Smith 1952, Hine 1969). Non-fatal incidents have occurred with estimated exposures possibly up to 192.8 mg/L for one to 20 min (Holling and Clarke 1944) and to 230 mg/L for 20 minutes (Henning 1933). The acute human exposure concentrations are speculative, due to the general nature of accidents. Exposure concentrations may be overestimated as methyl bromide is heavier than air and may be poorly mixed in the atmosphere. Other estimates of fatal and non-fatal exposure concentrations can be extrapolated from laboratory animal experiments.

VI. Delayed onset of symptoms in acute exposure cases

a) Animal

The time to death of laboratory animals exposed to methyl bromide, taken from the acute studies of Sayers *et al.* (1929) and Miyagawa *et al.* (1980), is presented in Table III. Examination of the Sayers *et al.* (1929) data indicates that for the same concentration, a shorter exposure usually resulted in a longer time before death. The Miyagawa *et al.* (1980) data indicated that for the same length of exposure, lower concentrations often resulted in a longer time before death. In both studies many of the animals died during exposure. In these cases it is difficult to determine whether the onset of death was delayed or whether the total exposure was required to produce death.

b) Human

A delay of more than an hour in the onset of toxic symptoms following methyl bromide exposure has been noted by several authors (Duvoir *et al.* 1937, Watrous *et al.* 1942, Holling and Clarke 1944, Wyers 1945, Andrieu *et al.* 1955, Longley and Jones 1965, Hine 1969). In these instances, the person was exposed to methyl bromide, but did not develop any symptoms for at least an hour after no longer being in contact with the compound. On the other hand, an equal number of instances indicate that effects from methyl bromide may be observed immediately, particularly at high concentrations (Duvoir *et al.* 1937, Billet and Abel 1940, Watrous 1942, Gray 1944, Holling and Clarke 1944, Wyers 1945, Pernod *et al.* 1961, Garcia Rico *et al.* 1974). By far the most common scenario is that the onset of the first symptoms occurred during acute exposure lasting one to 24 hr (Duvoir *et al.* 1937, Roger and Hawthorn 1941, Schifferli 1942, Watrous 1942, Miller 1943, Florentin 1944, Holling and Clarke 1944, Butler *et al.* 1945, Clarke *et al.* 1945, Thomson 1945, Viner 1945, Wyers 1945, Benatt and Courtney 1948, MacDonald *et al.* 1950, Irsigler 1951, Marchand-Aliphant 1952, Prain and Smith 1952, Jordi 1953, Hamon *et al.* 1955, Sigwald *et al.* 1955, Roche *et al.* 1958, Grossetette *et al.* 1963, Castaing *et al.* 1966, Goulon *et al.* 1975, Shield *et al.* 1977, Takayama and Numajiri 1979). Thus, in most cases it is difficult to differentiate between a delay in onset and lengthy, low-dose exposure that is required to produce toxicity. This concept should be kept in mind when considering the animal experiments of Sayers *et al.* (1929) and Irish *et al.* (1940), as well as the following information available on human exposure cases.

In several cases symptoms of eye irritation, throat irritation, dizziness, and vomiting have been reported immediately or within one hr following exposure (Duvoir *et al.* 1937, Billet and Abel 1940, Watrous 1942, Gray 1944, Holling and Clarke 1944, Clarke *et al.* 1945, MacDonald *et al.* 1950, Irsigler 1951, Marchand-Aliphant 1953, Jordi 1953, Garcia Rico *et al.* 1974). On occasion these symptoms were not reported until two hr (Wyers 1945, Longley and Jones 1965), four hr (Holling and Clarke 1944, Hine 1969), six hr (Duvoir *et al.* 1937), seven hr (Clarke *et al.* 1945), eight hr (Hine 1969), ten hr (Clarke *et al.* 1945), or 24 hr (Duvoir *et al.* 1937, Andrieu *et al.* 1955, Shield *et al.* 1977) following the initiation of exposure to methyl bromide. In some cases, the first symptoms noted were tremors, twitches, seizures, or convulsions. Of these, the symptom reported the soonest was hand twitching three hr after exposure (Duvoir *et al.* 1937). In one case tremors were reported 36 hr following initiation of exposure to methyl bromide (Mazel *et al.* 1946). Seizures have been reported at six hr (Miller 1943, Florentin 1944), 12 hr (Prain and Smith 1952), and 48 hr (Shield *et al.* 1977) following the initiation of exposure. The fact that these severe symptoms were the first symptoms noted in the case reports is probably due to the lack of observation of earlier symptoms, or the inability of the patients to report earlier symptoms. Severe poisoning cases that documented the case history from the time of exposure through the discharge of the patient generally indicated a progression from mild to moderate to severe symptoms. In such cases the mild symptoms developed one to six hr after exposure. The earliest that convulsions have been reported following exposure to methyl bromide is about four hr (MacDonald *et al.* 1950). Convulsions or seizures have also been reported at five hr (Holling and Clarke 1944), six hr (Irsigler 1951, Longley and Jones 1965), seven hr (Clarke *et al.* 1945), eight hr (Clarke *et al.* 1945), nine hr (Hine 1969), 13 hr (Mazel *et al.* 1946), 14 hr (Mazel *et al.* 1946, Prain and Smith 1952, Hine 1969), and about 27 hr (Duvoir *et al.* 1937) following the initiation of exposure to methyl bromide. In most severe but non-fatal acute exposure cases the complete progression of toxicity occurred within 14 hr of exposure. It seems that the symptoms developed more rapidly in the more severe exposure cases. Since exposure concentrations were not measured and the length of exposure is poorly documented, it is not possible to determine how closely the length and concentration of exposure correlates with the time to onset of symptoms.

In acutely fatal methyl bromide poisoning cases, the time from exposure to death varied from less than one hr (Garcia Rico *et al.* 1974) to 85 hr (Miller 1943). In a few cases death has occurred within ten hr (Holling and Clarke 1944, Wyers 1945, Mazel *et al.* 1946, Hine 1969). Most deaths from acute methyl bromide poisoning have occurred between 11 and 30 hr after exposure (Duvoir *et al.* 1937, Clarke *et al.* 1945, Mazel *et al.* 1946, MacDonald *et al.* 1950, Marchand-Aliphant 1952, Prain and Smith 1952, Jordi 1953, Hine 1969).

In cases of dermal exposure to methyl bromide, the onset of dermal irritation and blisters has ranged from one to 12 hr (De Gaulejac and Dervillee 1939, Thompson 1945, Wyers 1945, Benatt and Courtney 1948, Jordi 1953). The difference in time of onset may be due in part to the amount of clothing worn or the type of boots. The time to onset of systemic toxicity following dermal exposure does not appear to differ from the time to onset following inhalation exposure.

The time to onset of symptoms following an "acute" incident that was preceded by chronic methyl bromide exposure does not differ substantially from the time to onset after a single, unprecedented exposure. Generally the mild symptoms are reported while at work or shortly after completing work (Corcos *et al.* 1955, Franken *et al.* 1959, Rathus and Landy 1961, Williams and Reilly 1970, Greenberg 1971). In some cases ataxia is the first symptom reported at 12 to 14 hr following exposure to methyl bromide (Rathus and Landy 1961, Hine 1969).

The so-called chronic symptoms of methyl bromide poisoning such as twitching, tremors, ataxia, and mental disabilities have also developed on occasion following single acute exposures and "acute" exposures preceded by chronic exposure. In acute exposure cases, the "chronic" symptoms developed two to 15 days after exposure (Roger and Hawthorn 1941, Mazel *et al.* 1946, Irsigler 1951, Jordi 1953, Hamon *et al.* 1955, Andrieu *et al.* 1959, Longley and Jones 1965, Shield *et al.* 1977). In contrast, when acute exposure incidents were superimposed onto chronic exposures, the onset of "chronic" symptoms occurred 12 to 72 hr later (Duvoir *et al.* 1937, Rathus and Landy 1961, Williams and Reilly 1970, Greenberg 1971). Thus, the chronic exposure decreased the time to onset of prolonged toxic symptoms. In two of the acute cases the symptoms disappeared within a week. In the other cases the symptoms lasted from one mon to two yr (Duvoir *et al.* 1937, Roger and Hawthorn 1941, Irsigler 1951, Jordi 1953, Sigwald *et al.* 1955, Hamon *et al.* 1955, Andrieu *et al.* 1959, Bernot and Goldberg 1960, Rathus and Landy 1961, Kantarjian and Shaheen 1963, Langlois *et al.* 1963, Longley and Jones 1965, Castaing *et al.* 1966, Greenberg 1971, Garcia Rico *et al.* 1974).

VII. Acute non-neurological physiological effects

a) Animal

In the experiments of Sayers *et al.* (1929), the animals that died during the first 15 min of the exposure exhibited no non-neurological symptoms. But at necropsy, the pathological findings included dilation and degeneration of the heart, hemorrhage and consolidation of the lungs, and congestion and degeneration of the liver (gray in color), pancreas, and spleen. The animals that died 15 min to eight days following exposure exhibited coughing, retching-like action, increased respiration, and nasal discharges. The pathological findings of the animals dying within 24 hr were similar to the findings for animals that died within 15 min. For those that died from one to eight days after exposure, the spleen was pale or congested with some degenerated and hemorrhaged, and the suprarenals and intestines were congested. The researchers frequently noted blood-tinged intestinal contents. In non-fatal exposures, slight salivation and nasal discharge were reported. They noted in the controls slightly congested lungs, heart dilation, occasional kidney degeneration with a fatty cortex.

Irish *et al.* (1940) reported that at methyl bromide concentrations of 10 mg/L or greater, rats exhibited nose irritation, lachrymation, and heavy breathing. Necropsy showed the presence of pulmonary congestion, edema, possible bronchopneumonia, and kidney changes. Rabbits exposed to the same concentrations showed only slight lung irritation at necropsy. Exposure to concentrations of 1 to 10 mg/L caused rats to exhibit signs similar to higher exposures with occasional lachrymation and nasal irritation, while rabbits were not distinguishable from controls.

Miyagawa *et al.* (1980) reported that animals exposed to 40, 60, 100 or 120 ppm (for 24 hr) exhibited respiratory failure and salivation. Upon necropsy, the lungs were found to be hemorrhaged. The researchers noted a decrease in liver weight and increased testicular and adrenal weight.

b) Human

The non-neurological symptoms reported most often after acute inhalation of methyl bromide are effects on the respiratory system. Chest pain or difficulty in breathing are often reported along with pulmonary edema (Miller 1943, Holling and Clarke 1944, Clarke *et al.* 1945, Wyers 1945, Irsigler 1951, Marchand-Aliphant 1952, Derobert *et al.* 1953, Andrieu *et al.* 1955, Roche *et al.* 1958, Pernod *et al.* 1961, Grossetette *et al.* 1963, Castaing *et al.* 1966, Hine 1969, Williams and Reilly 1970, Shield *et al.* 1977, Maddy *et al.* 1981). These symptoms are consistent with the pathological findings at autopsy, which have included pulmonary edema, occasionally with blood or albuminous fluid (Duvoir *et al.* 1937, Miller 1943, Holling and Clarke 1944, Wyers 1945, Roche *et al.* 1958, Pernod *et al.* 1961, Mosinger *et al.* 1964, Hine 1969, Garcia Rico *et al.* 1974); bronchopneumonia (Viner 1945); congestion (Holling and Clarke 1945, Wyers 1945, Marchand-Aliphant 1952, Roche *et al.* 1958); and hemorrhaging

(Holling and Clarke 1944, Marchand-Aliphant 1952, Mosinger *et al.* 1964).

Other symptoms described after acute exposure include epigastric or abdominal pain, nausea or vomiting (Duvoir *et al.* 1937, Roger and Hawthorn 1941, Watrous 1942, Holling and Clark 1944, Wyers 1945, Irsigler 1951, Marchand-Aliphant 1952, Sigwald *et al.* 1955, Andrieu *et al.* 1959, Bernot and Goldberg 1960, Pernod *et al.* 1961, Castaing *et al.* 1966, Longley and Simpson 1969, Davay 1972, Garcia Rico *et al.* 1974, Bezuglyi *et al.* 1977, Shield *et al.* 1977, Maddy *et al.* 1981). In some cases pathological findings have indicated direct effects on the GI tract such as stomach congestion (Wyers 1945); submucous hemorrhage (MacDonald *et al.* 1950); and desquamation of surface cells of the stomach mucosa, small intestines, and large intestine (Miller 1943). It is not clear whether all these effects are direct or whether some are mediated via the central nervous system.

A dry throat and thirst have been reported in several cases (Miller 1943, Holling and Clarke 1944, Clarke *et al.* 1945, Wyers 1945, Marchand-Aliphant 1952, Sigwald *et al.* 1955, Shield *et al.* 1977, Maddy *et al.* 1981). Acute exposure victims have on occasion been described as "cyanotic" (Gray 1944, Holling and Clarke 1944, Wyers 1945, Thompson 1945, Prain and Smith 1952, Pernod *et al.* 1961, Castaing *et al.* 1966, Hine 1969, Garcia Rico *et al.* 1974).

Effects on the kidney have also been noted, particularly as anuria and blood or albumin in urine (Roger and Hawthorn 1941, Benatt and Courtney 1948, MacDonald *et al.* 1950, Marchand-Aliphant 1952, Prain and Smith 1952, Roche *et al.* 1958, Pernod *et al.* 1961, Grossetette *et al.* 1963, Langlois *et al.* 1963, Davay 1972, Goulon *et al.* 1975). Several pathological examinations at autopsy have suggested direct effects by methyl bromide on the kidney. The kidneys were reported to be congested (Wyers 1945, MacDonald *et al.* 1950, Roche *et al.* 1958, Mosinger *et al.* 1964), enlarged (Hine 1969), and inflamed (Viner 1945, Goulon *et al.* 1975). Close examination of the glomeruli revealed that they were on occasion dilated (Clarke *et al.* 1945, Mosinger *et al.* 1964) and congested (Holling and Clarke 1944, Marchand-Aliphant 1952). The tubules have exhibited degenerative changes (Holling and Clarke 1945, Roche *et al.* 1958, Pernod *et al.* 1961, Hine 1969) and necrosis (Duvoir *et al.* 1937, Holling and Clarke 1944, Clarke *et al.* 1945, Marchand-Aliphant 1952, Prain and Smith 1952).

The liver was found to be congested (Duvoir *et al.* 1937, Marchand-Aliphant 1952, Roche *et al.* 1958, Mosinger *et al.* 1964, Goulon *et al.* 1975) and fatty (Miller 1943, Holling and Clarke 1944). It is unlikely that this change is due to methyl bromide; it is probably due to ethanol consumption. Thus, an important question is the effect of chronic alcohol consumption on methyl bromide toxicity.

VIII. Subchronic and chronic non-neurological physiological effects

a) Animal

In the repeated exposure experiments of Irish *et al.* (1940), rats were exposed daily for approximately eight hr to 0.42 mg/L. They survived from one wk to more than five mon, but exhibited poor growth and appearance. Most of the animals exhibited normal lungs. Guinea pigs were more resistant at this concentration, had essentially normal growth with no observable pathological changes. Both rats and guinea pigs were indistinguishable from controls in terms of signs, growth, and histopathology following daily exposures to 0.25 mg/L of methyl bromide for six mon. In some cases at this concentration, the rabbits and monkeys showed signs of pulmonary congestion while all other tissues were normal. At the lowest exposure of 0.065 mg/L, growth, clinical signs, and histopathology were normal for all four species.

b) Human

The non-neurological symptoms reported following sub-chronic and chronic exposure to methyl bromide also include the respiratory tract or the gastrointestinal tract. Some general respiratory symptoms include chest pains, choking sensation, coughing, dyspnea, and pulmonary edema (De Gaulejac and Dervillee 1939, Johnstone 1945, Corcos *et al.* 1955, Franken *et al.* 1959, Kantarjian and Shaheen 1963, Hine 1969, Williams and Reilly 1970, Greenberg 1971, Ushio and Osozuka (1977). The fatal case autopsy reports of Franken *et al.* (1959) and Hine (1969) indicated pulmonary congestion. The gastrointestinal symptoms reported by persons suffering from chronic methyl bromide poisoning included abdominal pain, nausea, and vomiting (Duvoir *et al.* 1937, De Gaulejac and Dervillee 1939, Florentin 1944, Dejong 1944, Wyers 1945, Gayral 1949, Corcos *et al.* 1955, Franken *et al.* 1959, Rathus and Landy 1961, Drawneek *et al.* 1964, Hine 1969, Williams and Reilly 1970, Greenberg 1971, Araki *et al.* 1971, Yomura and Shibata 1978, Zatuchni and Hong 1981).

The only substantial differences in the non-neurological symptoms following acute and chronic exposures to methyl bromide are that some victims of acute toxic exposure developed dry throat, thirst, and eye irritation, with the kidneys affected on occasion. At the same time cyanosis has been reported in several acute cases while it was reported only in a few fatal chronic cases (Franken *et al.* 1959, Hine 1969).

IX. Acute neurological and behavioral manifestations

a) Animal

In the acute study of Sayers *et al.* (1929), neurological symptoms were observed in guinea pigs following most exposure concentrations. At 192.9 and 370.3 mg/L of methyl bromide, the guinea pigs quickly displayed uneasiness and lack of righting reflex. The slightly lower concentration of 84.9 mg/L induced gait incoordination and weakness. Exposure to 50.1 or 27.0 mg/L produced increased inactivity, gait incoordination, and weakness. The exposure concentrations of 7.7 to 8.9 mg/L produced weakness and eventual loss of righting reflex. At the lowest concentrations–0.58 and 0.39 mg/L–some animals eventually exhibited weakness.

Irish *et al.* (1940) reported in their acute experiments that methyl bromide exposure of 1.0 mg/L or greater to rats produce fur roughening, hunching, and drowsiness; these signs "usually occurred quickly." Rabbits exposed to 10 mg/L or greater generally sat with their eyes closed until narcosis. In exposure of 1.0 to 10 mg/L of methyl bromide, exposed rabbits were not distinguishable from controls. After 20 hr of exposure to 1.0 mg/L of methyl bromide, rats exhibited excitation and tremor while rabbits developed paralysis after two days.

Miyagawa *et al.* (1980) observed gait disturbances, paralysis, tremor and intermittent spasms in their 24-hr exposure experiments with rats. In a different study, Miyagawa *et al.* (1982) reported that rats exhibited increased taste aversion with increasing concentrations of methyl bromide. Homma *et al.* (1980) reported that exposure of rats to methyl bromide affected the concentrations of catecholamines, serontonin, acetylcholine, and amino acids in the corpus striatum, hypothalamus, and mid-brain.

b) Human

Schuler in 1899 described neurological features of methyl bromide poisoning. Yet due to idiosyncratic features of the poisoning syndrome and the limited experience of physicians with patients exposed to methyl bromide, subsequent cases have often been reported as new, with obscure neurological symptoms. This is reflected in some of the following titles of articles in the literature.

- "Chronic Epilepsy Following Methyl Bromide Poisoning" (Demousier and Steinmann 1936).
- "Methyl Bromide: An Insidious New Industrial Toxicant" (Florentin 1944).
- "Methyl Bromide Poisoning with Special Reference to Nervous System Manifestations" (Dejong 1944).
- "Methyl Bromide: A New Industrial Hazard" (Viner 1945).
- "Methyl Bromide Poisoning with Nervous System Manifestations Resembling Polyneuropathy" (Kantarjian and Shaheen 1963).
- "Methyl Bromide Poisoning: A Bizarre Neurological Disorder" (Collins 1965).
- "Methyl Bromide Intoxication: Neurological Features, Including Simulation of Reye Syndrome" (Shield *et al.* 1977).
- "Methyl Bromide Poisoning Seen Initially as Psychosis" (Zatuchni and Hong 1981).

The apprehension toward methyl bromide is understandable since the basis of knowledge concerning this compound is gross observations of poisoned human beings and exposed laboratory animals.

Numerous neurological and behavioral manifestations have been reported following acute inhalation or dermal exposure. Methyl bromide has produced some general effects on the central nervous system. The milder symptoms include headache, dizziness, and fainting (Duvoir *et al.* 1937, Billet and Abel 1940, Roger and Hawthorn 1941, Watrous 1942, Gray 1944, Holling and Clarke 1944, Wyers 1945, Thomson 1945, Benatt and Courtney 1948, MacDonald *et al.*

1950, Irsigler 1951, Marchand-Aliphant 1952, Jordi 1953, Hamon *et al.* 1955, Sigwald *et al.* 1955, Grossetette *et al.* 1963, Hine 1969, Longley and Simpson 1969, *San Francisco Chronicle* 1973, Garcia Rico *et al.* 1974, Bezuglyi *et al.* 1977, Shield *et al.* 1977, Maddy *et al.* 1981); drowsiness (Watrous 1942, Benatt and Courtney 1948, Pernod *et al.* 1961, Shield *et al.* 1977); apathy and lethargy (Holling and Clarke 1944, Benatt and Courtney 1948, Shield *et al.* 1977); and general weakness and tiredness (Watrous 1942, Holling and Clarke 1944, Thompson 1945, Jordi 1953, Andrieu *et al.* 1959, Castaing *et al.* 1966). Other general effects that have been reported are giddiness and delirium (Schifferli 1942, Holling and Clarke 1944, Wyers 1945, Grossetette *et al.* 1963, Shield *et al.* 1977); stupor (Duvoir *et al.* 1937, Roger and Hawthorn 1941, Holling and Clarke 1944, Jordi 1953, Andrieu *et al.* 1959, Pernod *et al.* 1961, Grossetette *et al.* 1963, Shield *et al.* 1977); psychosis and mood swings (Shield *et al.* 1977); loss of memory and mental confusion (Roger and Hawthorn 1941, Wyers 1945, Irsigler 1951, Jordi 1953, Hamon *et al.* 1955, Bernot and Goldberg 1960, Castaing *et al.* 1966, Garcia Rico *et al.* 1974, Shield *et al.* 1977); sweating (Thomson 1945, Pernod *et al.* 1961, Hine 1969); fever (Duvoir *et al.* 1937, Thomson 1945, Jordi 1953, Roche *et al.* 1958); anorexia (Watrous 1942); and vertigo (Watrous 1942, Garcia Rico *et al.* 1974, Maddy *et al.* 1981).

Acute exposure to methyl bromide has also been reported to cause more specific effects, such as slurred speech or other speech impairment (Cade and Mazel 1923, Roger and Hawthorn 1941, Florentin 1944, Wyers 1945, Irsigler 1951, Sigwald *et al.* 1955, Pernod *et al.* 1961, Langlois *et al.* 1963, Castaing *et al.* 1966, Shield *et al.* 1977, Ishizu *et al.* 1979); visual difficulty such as blurring, diplopia, and temporary blindness (Cade and Mazel 1923, Duvoir *et al.* 1937, Watrous 1942, Holling and Clarke 1944, Wyers 1945, Irsigler 1951, Pernod *et al.* 1961, Garcia Rico *et al.* 1974); and limb numbness, stiffness, weakness, pain, or cramps (Cade and Mazel 1923, Watrous 1942, Thomson 1945, Wyers 1945, Irsigler 1951, Longley and Jones 1965, Shield *et al.* 1977). Symptoms of greater severity affecting the limbs are tremors, muscle twitching (Roger and Hawthorn 1941, Florentin 1944, Wyers 1945, Irsigler 1951, Sigwald *et al.* 1955, Roche *et al.* 1958, Bernot and Goldberg 1960, Pernod *et al.* 1961, Grossetette *et al.* 1963, Langlois *et al.* 1963, Longley and Jones 1965, Castaing *et al.* 1966, Hine 1969, Garcia Rico *et al.* 1974, Bezuglyi *et al.* 1977, Takayama and Numajiri 1979), and paralysis (Cade and Mazel 1923, Jordi 1953, Takayama and Numajiri 1969).

In severe acute poisoning incidents, major neurological effects were observed, including ataxia (Roger and Hawthorn 1941, Bernot and Goldberg 1960, Pernod *et al.* 1961, Longley and Jones 1965, Shield *et al.* 1977); incoordination (Cade and Mazel 1923, Florentin 1944, Wyers 1945, Irsigler 1951, Sigwald *et al.* 1955, Bernot and Goldberg 1960, Pernod *et al.* 1961, Garcia Rico *et al.* 1974); Jacksonian seizures (Andrieu *et al.* 1959, Longley and Jones 1965); status epilepticus, epileptiform convulsions, and clonic-tonic seizures (Duvoir *et al.* 1937, Roger and Hawthorn 1941, Florentin 1944, Holling and Clarke 1944, Clarke *et al.* 1945, Viner 1945, Wyers 1945, MacDonald *et al.* 1950, Irsigler 1951, Marchand-

Aliphant 1952, Hamon *et al.* 1955, Sigwald *et al.* 1955, Roche *et al.* 1958, Grossetette *et al.* 1963, Longley and Jones 1965, Castaing *et al.* 1966, Hine 1969, Garcia Rico *et al.* 1974, Mellerio *et al.* 1974, Davay 1974); and narcosis or unconsciousness (Duvoir *et al.* 1937, Roger and Hawthorn 1941, Florentin 1944, Holling and Clarke 1944, Viner 1945, Wyers 1945, MacDonald *et al.* 1950, Irsigler 1951, Marchand-Aliphant 1952, Jordi 1953, Hamon *et al.* 1955, Sigwald *et al.* 1955, Roche *et al.* 1958, Andrieu *et al.* 1959, Pernod *et al.* 1961, Grossetette *et al.* 1963, Langlois *et al.* 1963, Longley and Jones 1965, Castaing *et al.* 1966, Hine 1969, Garcia Rico *et al.* 1974, Mellerio *et al.* 1974, Davay 1974, Bezuglyi *et al.* 1977, Ishizu *et al.* 1979).

Examination of the brain at autopsy has primarily revealed congestion with multiple minute hemorrhages (Duvoir *et al.* 1937, Holling and Clarke 1944, Clarke *et al.* 1945, Viner 1945, Wyers 1945, MacDonald *et al.* 1950, Prain and Smith 1952, Jordi 1953, Roche *et al.* 1958, Pernod *et al.* 1961, Mosinger *et al.* 1964, Garcia Rico *et al.* 1974), and spongy necrosis and demyelination of dentate nuclei (Girard *et al.* 1960). Goulon *et al.* (1975) reported moderate changes in the dentate and pontine nuclei, gliosis in the upper brainstem, reticular formation, and necrosis of both inferior coliculi.

X. Subchronic and chronic neurological and behavioral manifestations

a) Animal

In the chronic experiments of Irish *et al.* (1940), rats exhibited occasional convulsions following exposures of 0.42 mg/L. At 0.25 mg/L rabbits developed paralysis after 14 exposures while monkeys occasionally developed paralysis after three mon of exposure. The paralysis disappeared following removal from the exposure chamber. At the exposure concentrations of 0.13 mg/L, only rabbits exhibited paralysis, with recovery on removal from the exposure chamber. The gross examination of the brain and spinal cord of several paralyzed rabbits and monkeys did not reveal any congestion, edema, or hypertrophy.

Ikeda *et al.* (1980) studied the effects of a three-wk exposure of rats to methyl bromide. The 200 ppm group was asymptomatic while the 300 ppm groups exhibited depressed spontaneous activity and one animal had convulsions. Both groups had a significantly lower weight gain than controls ($p < 0.01$) until 25 days after the last exposure. From rotored tests conducted 12 days after the final exposure they reported a decrease in performance in the 300 ppm exposure group, and from tests at 28 days, a decrease in performance in both exposure groups. The authors reported disturbance of the circadian rhythm at the 300 ppm exposure level. Upon necropsy, no macroscopic or microscopic changes were noted in the central nervous system or peripheral nerves.

Anger *et al.* (1981) repeatedly exposed rabbits to 65 ppm methyl bromide for one mon. A weight loss was observed, but it was not statistically significant. They found a significant decrease in sciatic and ulnar nerve impulse velocities in rabbits. An eyeblink reflext test indicated a decrease in mean eyeblink amplitude over time during exposure, but the final amplitudes were not significantly different from controls (p=0.08, t-test). No significant changes in sciatic nerve or ulnar nerve impulse velocities were determined in rats exposed to 65 or 55 ppm of methyl bromide for either one mon or one yr (respectively).

b) Human

Some of the mild signs of central nervous system (CNS) involvement in subchronic and chronic methyl bromide toxicity are headache, dizziness, and fainting (Duvoir *et al.* 1937, De Gaulejac and Dervillee 1939, Dejong 1944, Florentin 1944, Carter 1945, Johnstone 1945, Viner 1945, Gayral 1949, Corcos *et al.* 1955, Rathus and Landy 1961, Kantarjian and Shaheen 1963, Drawneek *et al.* 1964, Hine 1969, Araki *et al.* 1971, Greenberg 1971, Takahashi *et al.* 1973, Yoshida *et al.* 1975, Ushio and Osozuka 1977, Yomura and Shibata 1978, Zatuchni and Hong 1981) and weakness, fatigue, tiredness, dullness, and depression (De Gaulejac and Dervillee 1939, Dejong 1944, Carter 1945, Johnstone 1945, Wyers 1945, Gayral 1949, Corcos *et al.* 1955, Franken *et al.* 1959, Kantarjian and Shaheen 1963, Drawneek *et al.* 1964, Collins 1965, Hine 1969, Ushio and Osozuka 1977, Yomura and Shibata 1978, Zatuchni and Hong 1981). Many other CNS-related symptoms have been ascribed to chronic methyl bromide poisoning, including loss of initiative, loss of libido, inability to tolerate beer, hallucinations, confusion, euphoria, personality changes, irritability, psychosis, and neurosis.

In subchronic or chronic poisoning cases some of the earlier symptoms include slurring of speech and speech impairment (Duvoir *et al.* 1937, Carter 1945, Johnstone 1945, Viner 1945, Wyers 1945, Eckardt 1970, Greenberg 1971); diplopia, blackouts, or impaired vision (Duvoir *et al.* 1937, Florentin 1944, Dejong 1944, Carter 1945, Johnstone 1945, Wyers 1945, Gayral 1949, Rathus and Landy 1961, Drawneek *et al.* 1964, Hine 1969, Araki *et al.* 1975, Yomura and Shibata 1978); tinnitus or other auditory disturbances (Dejong 1944, Rathus and Landy 1961, Hine 1969); anorexia (Duvoir *et al.* 1937, De Gaulejac and Dervillee 1939, Dejong 1944, Viner 1945, Franken *et al.* 1959, Kantarjian and Shaheen 1963, Hine 1969, Ushio and Osozuka 1977, Zatuchni and Hong 1981); and vertigo (Duvoir *et al.* 1937, Dejong 1944, Gayral 1949, Drawneek *et al.* 1964, Hine 1969). Victims occasionally complain of numbness or tingling in the extremities (Dejong 1944, Carter 1945, Viner 1945, Johnstone 1945, Kantarjian and Shaheen 1963, Takahashi *et al.* 1973).

Of greater concern is the development of a lack of coordination (Duvoir *et al.* 1937, Florentin 1944, Carter 1945, Johnstone 1945, Viner 1945, Gayral 1949, Rathus and Landy 1961, Kantarjian and Shaheen 1963, Drawneek *et al.* 1964, Collins 1965, Hine 1969, Eckardt 1970, Greenberg 1971) and ataxia or gait disturbance (Duvoir *et al.* 1937, Dejong 1944, Carter 1945, Viner 1945,

Gayral 1949, Rathus and Landy 1961, Kantarjian and Shaheen 1963, Drawneek *et al.* 1964, Collins 1965, Williams and Reilly 1970, Eckardt 1970, Araki *et al.* 1971, Greenberg 1971, Takahashi *et al.* 1973, Ushio and Osozuka 1977, Yomura and Shibata 1978, Zatuchni and Hong 1981). In severe cases, muscular tremors and twitches (Florentin 1944, Johnstone 1945, Viner 1945, Franken *et al.* 1959, Rathus and Landy 1961, Drawneek *et al.* 1964, Greenberg 1971, Takahashi 1973), epileptiform seizures or Jacksonian convulsions (Florentin 1944, Johnstone 1945, Corcos *et al.* 1955, Franken *et al.* 1959, Rathus and Landy 1961, Hine 1969, Greenberg 1971), unconsciousness (Johnstone 1945, Corcos *et al.* 1955, Franken *et al.* 1959, Rathus and Landy 1961, Hine 1969, Williams and Reilly 1970, Greenberg 1971, Ushio and Osozuka 1977), and loss of memory (De Gaulejac and Dervillee 1939, Johnstone 1945, Wyers 1945, Hine 1969, Greenberg 1971) were reported.

The convulsive and narcotic actions of methyl bromide are usually associated with acute exposure. These symptoms were also reported in chronic exposure cases, but the case descriptions indicated that a larger than usual exposure incident initiated the symptoms (Johnstone 1945, Corcos *et al.* 1955, Franken *et al.* 1959, Rathus and Landy 1961, Hine 1969, Williams and Reilly 1970, Greenberg 1971, Ushio and Osozuka 1977). As several of the above authors have noted, it is possible that chronic exposure resulted in carelessness in handling the methyl bromide, increasing the likelihood of a spillage or leakage accident. Further, the chronic exposure may have decreased the threshold for acute exposure toxicity. Thus, a single exposure to methyl bromide that would not precipitate an acute crisis in a previously unexposed person, may produce "acute" symptoms, such as convulsions, in an individual who had been chronically exposed, such as a fumigator.

In a case described by Hine (1969), a laborer had a large exposure superimposed on his chronic exposure and the resulting autopsy indicated the brain was swollen and edematous. In a similar case, Franken *et al.* (1959) noted lesions of the motor spinal roots ganglia, demyelination of the cerebellar peduncles, and lesions in the cerebral and cerebellar cortices, but the authors concluded that it was not possible to distinguish between the effects of the methyl bromide and the terminal seizures themselves.

A typical acute exposure syndrome begins with the development of a headache or dizziness accompanied by gastrointestinal and respiratory symptoms. This is followed by muscular pain, numbness, or twitching in the extremities. In severe cases, this is followed by convulsions and unconsciousness. Upon initial recovery in severe non-fatal cases, there is a chance of developing some long-lasting symptoms affecting vision, speech, memory, or muscular coordination.

At the onset of a typical chronic case there may be headache, dizziness, gastrointestinal symptoms, anorexia, speech impairment, visual problems, or general weakness. In more severe cases there is a development of motor incoordination and ataxia. If an "acute" exposure is superimposed on a chronic exposure, there is a likelihood of convulsions and unconsciousness.

XI. Mutagenicity and cytotoxicity

The mutagenicity of methyl bromide was tested by Simmon *et al.* (1977 and 1978) in *Salmonella typhinurium* TA 100 using the Ames assay. Due to methyl bromide's volatility, the compound was introduced into a nine-L desiccator at concentrations from 0.39 to 7.71 mg/L. The researchers found a dose-dependent increase in the number of mutants in the absence of metabolic activation. They reported that the mutagenicity increased with chemical reactivity, with the following order of mutagenicity: methyl iodide > methyl bromide > methyl chloride, and methyl bromide > bromochloromethane > methylene chloride.

The mutagenicity of methyl bromide was determined in barley kernels by Ehrenberg *et al.* (1974). They found that at the LC_{50}, mutation frequency of methyl bromide was less than one-tenth that of methyl methanesulfonate and ethylene dibromide. When comparing the compounds' reactivity, the lethal effectiveness and mutagenic effectiveness of methyl bromide were equivalent to that of methyl methanesulfonate, but methyl bromide's lethal effectiveness was less than 1/50 of ethylene dibromide's while its mutagenic effectiveness was less than 1/800 of ethylene dibromide's. Thus, methyl bromide exhibited some mutagenic activity but considerably less than ethylene dibromide.

Djalali-Behzad *et al.* (1981) studied the mutagenicity of methyl bromide in an *E. coli* forward mutation system. After incubation of bacteria in a solution of 5,000 mg/L methyl bromide, the mutagenic effectiveness was estimated at one mutation/100,000,000 surviving bacteria. The mutagenic efficiency was low due to the compound's high cytotoxicity. They demonstrated guanine-N-7 alkylation.

Nishimura *et al.* (1980) examined the cytotoxic effects of methyl bromide on mammalian cells. They found that all cells died after a three-day exposure to a 10 mg/L solution. At lower concentrations, pyknosis increased with exposures up to 24 hr, but did not continue to increase with further exposures up to 72 hr. Delayed cytotoxic effects were not observed.

XII. Teratogenicity

Hardin *et al.* (1981) exposed rats and rabbits to 20 or 70 ppm of methyl bromide for seven hr/day, five days/wk for three wk. The rats were mated and exposed on days 1 to 19 of gestation. At these concentrations and exposures, methyl bromide did not produce any maternal toxicity, fetal toxicity, or teratogenicity. The rabbits were mated and exposed only on days 1 to 15 of gestation due to the development of maternal neurotoxicity. At the 70 ppm concentration, 24 of 25 rabbits died. At the 20 ppm level the researchers did not observe any maternal toxicity, fetal toxicity, or teratogenicity.

XIII. Carcinogenicity

No studies have yet been conducted on the carcinogenicity of methyl bromide in laboratory animals. As indicated by Lawley (1976), the repeated exposure experiments of Irish *et al.* (1940) did not produce any detectable tumors, but the results are probably inconclusive. *In vitro* studies do indicate that the compound is mutagenic. Methyl bromide is a well-known alkylating agent as discussed below in Section XIV. Related compounds that have been demonstrated to have some carcinogenic potential are methyl chloride (Pavkov *et al.* 1982) and methyl iodide (reviewed by Lawley 1976).

XIV. Biochemical fate

A number of experiments have been conducted regarding the nature of methyl bromide's decomposition when in the presence of biological constituents. The rate of decomposition of methyl bromide during fumigation of wheat flour was greater in low-moisture samples, which led Winteringham and Harrison (1946) to conclude that simple hydrolysis was not the primary method of decomposition. This was confirmed in a later experiment (Winteringham *et al.* 1955), in which they noted that moist gluten was less reactive with methyl bromide than dry gluten and that methanol production (*e.g.*, via hydrolysis) accounted for less than 10% of the reaction products.

The bulk of investigations conducted support alkylation as the method of methyl bromide decomposition. Lewis and Eccleston (1946) reported that, after fumigating wheat flour with 80 mg/L of methyl bromide for 48 hr, the amount of bromide ion residue detected was proportional to the product's protein content. They suggested that the protein fraction was primarily responsible for the decomposition of methyl bromide. Winteringham and Harrison (1946) found, after fumigating wheat flour with methyl bromide concentrations ranging from 20 to 135 mg/L, that the protein fraction was responsible for the physical adsorption of methyl bromide.

The sulfhydryl group has been implicated as a primary site for methyl bromide alkylation. Blackburn *et al.* (1944) placed bisulfited wool in a stoppered bottle of methyl bromide for three days. Analysis showed that thiol groups were methylated. Toennies and Kolb (1945) showed that methyl bromide can react with methionine *in vitro* to form alkyl sulfonium salts. Dixon and Needham (1946) reported that methyl bromide inhibited hexokinase *in vitro.* They suggested that the inhibition occurred due to reaction with the thiol groups. Lewis (1948) exposed solutions of cysteine and glutathione to vapor concentrations of methyl bromide from 98 to 200 mg/L for up to four hr. He found that the sulfhydryl methylation was directly proportional to the amount of bromide ion detected. When reduced keratin was methylated by a concentration of 400 mg/L for two hr, the nitroprusside reaction was abolished. The final phase of the study showed enzymatic inhibition of urease, succinic dehydrogenase, and papain with solution concentrations of 1,130, 1,320, and 510 mg/L of methyl bromide, respectively. Lewis hypothesized that the toxicity of methyl bromide may arise from its reactions with protein sulfhydryl groups.

The alkylation of sulfhydryl groups was indirectly shown *in vitro* by Nishimura *et al.* (1980). Cytotoxicity was decreased by the addition of reduced glutathione, suggesting that alkylation of the glutathione diminished the toxic effects of methyl bromide.

A study by Starratt and Bond (1981) compared the glutathione levels in resistant and susceptible granary weevils and found that those resistant to methyl bromide (requiring twice the expected dose) had higher glutathione levels. The two major metabolites they obtained following radiolabeled methyl bromide exposure were *S*-methyl glutathione and *S*-methyl cysteine. Susceptible strains had a higher proportion of total radioactivity which was insoluble in 60% ethanol.

Aside from reacting with sulfhydryl groups, methyl bromide can undergo nucleophilic substitution with other substrates. Blackburn *et al.* (1941) methylated wool and found that the primary alkylation reactions that occurred were *O*- and *N*-methylation. In a related study, Blackburn *et al.* (1944) found that collagen and gelatin were methylated with methyl bromide. The primary nucleophilic substitution reaction appeared to be esterification of free carboxyl groups. Winteringham *et al.* (1955) investigated the fate of radiolabelled methyl bromide by the fumigation of wheat flour with 153 mg/L of $^{14}CH_3Br$ for 41 hr. They found that roughly 80% of the ^{14}C-label was present in the protein fraction. Fifty % of the compounds in this fraction were *N*-methylated, 40% were *S*-methylated, and 10% were *O*-methylated.

To determine the nature of the *N*-methylated derivatives following radiolabelled methyl bromide fumigation of flour, Bridges (1955) analyzed the compounds using potato-starch chromatography. The *N*-methylhistidine derivatives 1-*N*-methylhistidine, ε-*N*-methylhistidine, and 1,3-*N*-methylhistidine accounted for roughly 75% of the total *N*-methylation while ε-*N*-methylated lysine accounted for 10%. Dunkelburg (1980) methylated the amino acids glycine, alanine, valine, leucine, and glutamic acid in an aqueous solution of 80 mg/L methyl bromide. In each case an *N*-methyl derivative was isolated.

The alkylating ability of methyl bromide *in vitro* and *in vivo* was investigated by Djalali-Behzad *et al.* (1981). In a set of *in vitro* experiments with ^{14}C-labelled methyl bromide, DNA was incubated in a 1.08 mg/L solution for 60 min, mouse spleen cells were incubated in a 1.57 mg/L solution for 34 min, and mouse red blood cells were incubated in a 0.36 mg/L solution for 60 min. The researchers determined the alkylation rate constants for guanine-*N*-7 in the DNA solution and suspended spleen cells, histidine in the suspended erythrocytes and spleen cells, and cysteine-*S* in suspended erythrocytes. The degree of DNA alkylation in the *in vitro* experiments was compared to that of *in vivo* alkylation in mice. Mice were given ^{14}C-labelled methyl bromide by i.p. at 0.42 mg/kg or by inhalation at either 3.35 or 6.55 mg/kg (calculated after ^{14}C-measurement). From the determined alkylation reactivity of the DNA in solution, the degree of DNA alkylation in the spleen was only 5% of the expected *in vivo* and 12% *in vitro*. The degree of DNA alkylation in the liver *in vivo* was one-tenth that of the spleen. There was also an unexpectedly low degree of liver protein alkylation.

XV. Toxicokinetics

Methyl bromide is absorbed through the lungs, gastrointestinal tract, and probably through the skin. The pulmonary absorption of methyl bromide is evident from human exposure cases and has been the primary route of exposure for laboratory animals. The gastrointestinal absorption of methyl bromide has been demonstrated by Irish *et al.* (1941), Dudley and Neal (1942), and Miller and Haggard (1943). No dermal experiments using laboratory animals have been reported; belief in dermal absorption is based on the systemic effects noted in human poisoning cases where methyl bromide came in contact with the skin, as described above. The compound is slightly soluble in water and very soluble in alcohol and ether (Weast 1971). These properties indicate that dermal absorption is likely. Since the commercial advantage of methyl bromide is its ability to penetrate food commodities, the compound should be absorbed through human and animal tissue. It is difficult to eliminate absolutely the possibility of inhalation exposure in the cases attributed to dermal absorption. The absence of systemic symptoms in some cases where severe dermal irritation is evident confuses the issue further. Although the dermal absorption of methyl bromide has not been clearly demonstrated, it is suspected and should be studied further.

The uptake of methyl bromide by rats was studied indirectly by Anderson *et al.* (1980) using a closed-system exposure chamber and measuring the decrease in air concentration over time. They used methyl bromide concentrations of 10,000, 3,000, 1,000, and 100 ppm. The animals were exposed for up to three hr. For concentrations up to 3,000 ppm, the uptake of methyl bromide was first-order. At the highest concentration, the animals died during the exposure and the apparent first-order rate constant decreased. Anderson *et al.* (1980) found that the uptake of methyl bromide was related to the production of bromide ion and that the high uptake rate implicates a nonenzymatic metabolism. The high rate of metabolism to bromide ion was later confirmed by Gargas *et al.* (1980). The conversion of methyl bromide to bromide ion has been shown in numerous human case reports and animal exposure experiments, and is described below in Section XVII.

The solubility of methyl bromide indicates that it should be distributed to essentially all tissues. The wide tissue distribution of bromide ion, including abdominal fat, following fatal exposure to methyl bromide in two dogs by Duvoir *et al.* (1939) indirectly supports this hypothesis. Sato *et al.* (1980 a) exposed rats for up to four hr to 400 ppm of methyl bromide. They found that the tissue methyl bromide concentrations in blood, liver, kidney, heart, and brain reached a maximum after one hr of exposure and were undetectable one to two hr following termination of the exposure. They also found that methyl bromide metabolism in the liver was relatively rapid, but that tissue bromide concentrations reached a maximum one to three days after exposure.

The half-life of methyl bromide has not been reported, but the half-life of the bromide ion has been reported to be 12 days in humans (Soremark 1960 a), 3.5 days in rats (Rauws and Van Logten 1974), and 1.5 days in mice (Soremark 1960 b). The pharmacokinetics of bromide ion was discussed by Rauws (1975).

XVI. Mechanism of action

The mechanism of toxic action of methyl bromide has not been determined. Miller and Haggard (1943) dosed rats orally and intraperitoneally with 75 to 180 mg/kg of methyl bromide in olive oil. They reported that mortality was related to bromide ion concentrations in the animal. They further indicated that a larger proportion of the bromide ion was located intracellularly following methyl bromide administration as compared with that of sodium bromide. Their conclusion was that methyl bromide toxicity may be due to "intracellular brominism." Although they assumed that the method of methyl bromide decomposition was hydrolysis, the potential toxicity of intracellular bromide ion is not dependent on the mode of decomposition. The theory of "intracellular brominism" has not received substantial support.

An experiment by Nishimura *et al.* (1980) considered the effects of methyl brominated bovine serum on HeLa cells. No cytotoxicity was observed, indicating that reaction products are not likely to be causative agents. However, the Miller and Haggard hypothesis is not refuted, since the HeLa cell exposure was extracellular.

The possibility of methyl alcohol as the actual cause is unlikely since it is less toxic than methyl bromide itself and does not produce a similar poisoning syndrome in humans. Another suggested mechanism, related to the lipid solubility of methyl bromide, is the direct disruption of cellular membranes, particularly nerve cells. Experiments designed to test this hypothesis could be fruitful.

By far the greatest interest has been placed in hypotheses that relate to the alkylating ability of methyl bromide. The alkylation of sulfhydryl groups in crucial enzymes has been suggested by Lewis (1948) and Winteringham and Barnes (1955). As indicated above, the alkylation of such groups has been strongly implicated *in vitro* and *in vivo*. But methyl bromide alkylation is not restricted to sulfhydryl groups (Blackburn *et al.* 1941 and 1944, Winteringham 1955, Bridges 1955, Dunkelburg 1980, Djalali-Behzad *et al.* 1981). The greatest difficulty in ascertaining the mechanism of action of methyl bromide relates to this lack of substrate specificity. If certain associations of substrate reactivity and toxicity are determined, they should not be assumed to be the molecular target causing toxicity. In any case, a coherent theory on toxicity should take into consideration the solubility properties of methyl bromide and the sensitivity of neurons.

The potential toxicologic and nutritional significance of methylation by methyl bromide was reviewed by Winteringham (1955). He reported that the metabolites produced were not considered significantly toxic to mammals to account for methyl bromide's observed toxicity. Due to the low percentage of total methionine and histidine converted following methylation, no nutrition effects were anticipated. In a study by Friedman and Gumbmann (1979), mice were fed with methylated lysine and histidine instead of the unmethylated compounds. When either of these amino acids was substituted, the mean weight gain was significantly less than that of the controls.

XVII. Diagnosis and detection of exposure

a) Animal

Bromide ion measurements in tissues have not generally been reported in animal studies of methyl bromide exposure. Irish *et al.* (1941) reported that control rabbits had blood bromide values of ten ppm, while those exhibiting toxicity from chronic inhalation exposure to 0.23 mg/L had bromide ion levels of 110 ppm. Repeatedly dosing rabbits for 22 days with 30 mg/kg of methyl bromide produced similar toxic symptoms and blood bromide ion levels of 200 ppm. A study by Uchiyama *et al.* (1979) exposed groups of rabbits to three consecutive exposures of 50, 300 or 500 ppm of methyl bromide and found maximum blood bromide ion levels of 60, 200, and 200 ppm, respectively. Sato *et al.* (1980 b) exposed rats to 200, 300 or 400 ppm of methyl bromide for four hr/day, five days/wk for six wk. After sacrifice, they detected bromide ion in the liver at the three dose levels to be 15, 15, and 70 ppm, respectively. For all groups, they reported 100 ppm of bromide in the kidney and 80 ppm of bromide in the spleen.

Studies have been conducted in which animals were fed fumigated diets and tissue bromide ion levels were measured (Spencer *et al.* 1944, Getzendaner 1965, Williford *et al.* 1974, Yomura *et al.* 1976). Other experiments were conducted in which animals were fed diets with sodium bromide ion added and the bromide ion tissue levels were measured (Van Logten *et al.* 1973, Knight and Reina-Guerra 1977). These studies have shown that bromide ion intake is correlated with tissue concentrations and excretion, and that toxic symptoms develop when tissue bromide ion levels reach approximately 500 to 2,000 ppm. The signs exhibited are unlike those associated with methyl bromide poisoning.

Several laboratory animal experiments have considered the therapeutic value of British anti-lewisite (BAL), cysteine, and acetylcysteine. Winteringham and Barnes (1955) noted that BAL could protect rats from methyl bromide poisoning if it was given prophylactically, but not after exposure. Kawai and Ueda (1972) administered either glutathione or BAL to rats prior to a one-hr 7.14

mg/L methyl bromide exposure. Of the controls, 100% died, but the percent mortality for those administered 750 mg/kg GSH, 250 mg/kg GSH and 50 mg/kg BAL (all i.p.) were 0, 60, and 60, respectively. Bakhishev (1978) reported that methyl bromide toxicity in rats was decreased following the administration of acetylcysteine, glutathione, and cysteine salts but not with BAL. Following exposure to 12.5 mg/L of methyl bromide (the 30 min LC_{99}), cysteine was administered at 150 mg/kg every 30 min until observable symptoms and 98% of the rats survived. When cysteine was administered (at 500 or 750 mg/kg) 30 min prior to exposure, 87% of the rats survived. And when cysteine was administered in three doses (300, 200, and 100 mg/kg) following poisoning of the animals, only 75% survived.

The study on the granary weevil by Starratt and Bond (1981) indicates that resistant strains have greater glutathione conjugation with methyl bromide following exposure. The study of Nishimura *et al.* (1980) found that the addition of glutathione reduced the cytotoxicity of methyl bromide in mammalian cell cultures.

b) Human

The diagnosis of methyl bromide poisoning is based on the patient's exposure history, medical examinations, and symptoms (Holling and Clarke 1944, Ingram 1951, Drawneek *et al.* 1964). Without knowledge of the exposure history, a poisoning incident may be mistaken for another disease or syndrome (Holling and Clarke 1944, Clarke *et al.* 1945, MacDonald *et al.* 1950, Collins 1965, Eckardt 1971, Shield *et al.* 1977, Zatuchni and Hong 1981). Since methyl bromide poisoning symptoms may not appear for several hr following an incident, the compound may not be immediately suspected. Although there are numerous case reports in the literature describing the toxic effects of methyl bromide following inhalation, it is difficult to relate observable symptoms with exposure. The symptoms and signs recorded depend in part on the presiding clinician.

For diagnostic purposes bromide ion levels have often been measured. Most of the bromide ion values reported in human exposure cases have been determined in either whole blood or serum. These values are presented in Table IV. Following acute exposure to methyl bromide, the serum bromide ion half-life can be calculated to be approximately 8.5 days in a patient demonstrating kidney dysfunction (Benatt and Courtney 1948). The half-lives of bromide ion that can be calculated in patients hospitalized following chronic intoxication of methyl bromide ranged from 3.5 to 15 days. The variation may be related to chloride intake, since Rauws and Van Logten (1975) showed that bromide excretion was increased in rats with high chloride intake. The half-life in human beings for bromide ion, following administration of ^{82}Br, is 12 days (Soremark 1960 a).

The severity of symptoms was related to serum and blood bromide ion values obtained from the literature (Table IV). The bromide ion levels measured within three days following methyl bromide exposure were used. The mean bromide

Table IV. *Blood and serum bromide ion levels in human methyl bromide exposure cases.*

A. Unexposed controls

Blood Br^- (ppm)	Serum Br^- (ppm)	Reference
0.1-3.5	–	Ucko (1936)
0.5-2.0	–	Hine (1969)
3.7	–	Conway & Flood (1936)
7.3		Dixon (1935)
6-8	–	Rathus (1980)
8.1	–	Wikoff *et al.* (1940)
10.0	–	Zondek & Bier (1931)
15	–	Clarke *et al.* (1945)
–	0-15	Benatt & Courtney (1948)
–	14	Drawneek *et al.* (1964)
–	30	Zatuchni & Hong (1981)

B. Acute exposure cases

Mild to moderate symptoms

Days after exposure	Blood Br^- (ppm)	Serum Br^- (ppm)	Reference
1	69	–	Holling & Clarke (1944)
1	69	–	Clarke *et al.* (1945)
1	69	–	Clarke *et al.* (1945)
4	–	125	Benatt & Courtney (1948)
7	–	134	Benatt & Courtney (1948)
10	–	70	Benatt & Courtney (1948)
12	69	–	Holling & Clarke (1944)
13	69	–	Clarke *et al.* (1944)
13	69	–	Clarke *et al.* (1944)
13	–	53	Benatt & Courtney (1948)
19	–	46	Benatt & Courtney (1948)
32	8	–	Clarke *et al.* (1944)
40	0	–	Clarke *et al.* (1944)
42	0	–	Clarke *et al.* (1944)

Severe symptoms and fatalities

Days after exposure	Blood Br^- (ppm)	Serum Br^- (ppm)	Reference
0.5	400	–	Hine (1969)
0.5	300	–	MacDonald *et al.* (1950)
1	–	550	Longley & Jones (1965)
1	92	–	Clarke *et al.* (1945)
1	83	–	Clarke *et al.* (1945)

Table IV (*continued*)

Days after exposure	Blood Br− (ppm)	Serum Br− (ppm)	Reference
1	50[a]	–	Hine (1969)
1.5	93	–	Holling & Clarke (1944)
1.5	83	–	Holling & Clarke (1944)
1-2	–	445	Takayama & Numajiri (1970)
5	–	2116	Viner (1945)

C. Chronic exposure cases

No symptoms

Blood Br^- (ppm)	Serum Br^- (ppm)	Reference
32, 36, 12, 8, 12, 5, 16, 12, 12, 4, 4, 18		Rathus & Landy (1961)
	0, 0, 9, 36, 54, 64, 88, 73, 128, 118, 131, 123, 136, 153	Drawneek *et al.* (1964)
4, 4.3, 4.7, 5, 5, 5, 6, 6.2, 6.7, 6.7, 7, 7, 7.3, 8.3, 8.3, 8.5, 9, 10, 11, 12, 12.3, 12.3, 18, 4.5, 5.1, 5.2, 8.2, 12, 12.2, 16.2, 17, 18, 23		Verberk *et al.* (1979)

Mild to moderate symptoms

Days after last exposure	Blood Br^- (ppm)	Serum Br^- (ppm)	Reference
1	150	–	Hine (1969)
1	120	–	Hine (1969)
1	96	–	Rathus & Landy (1961)
1	80	–	Rathus & Landy (1961)
1	55	–	Rathus & Landy (1961)
1	–	187	Drawneek *et al.* (1964)
1	–	120	Zatuchni & Hong (1981)
1-2	130	–	Dejong (1944)
2	220	–	Hine (1969)
3	64	–	Rathus & Landy (1961)

[a]Possible error per author.

Table IV (*continued*)

Days after last exposure	Blood Br− (ppm)	Serum Br− (ppm)	Reference
6	–	105	Drawneek *et al.* (1964)
6	–	50	Zatuchni & Hong (1981)
12	–	104	Drawneek *et al.* (1964)
12	35	–	Rathus & Landy (1961)
14	15	–	Rathus & Landy (1961)
18	24	–	Rathus & Landy (1961)
30	24	–	Rathus & Landy (1961)
30	5	–	Rathus & Landy (1961)
60	15	–	Rathus & Landy (1961)
60	0	–	Rathus & Landy (1961)
90	–	104	Drawneek *et al.* (1964)
98	0	–	Rathus & Landy (1961)
131	0	–	Rathus & Landy (1961)
150	–	65	Drawneek *et al.* (1964)

Severe symptoms and fatalities

Days after last exposure	Blood Br^- (ppm)	Serum Br^- (ppm)	Reference
1	400	–	Rathus & Landy (1961)
1	250	–	Rathus & Landy (1961)
1	135	–	Rathus & Landy (1961)
1	72	–	Rathus & Landy (1961)
1	–	500	Viner (1945)
1	–	328	Ushio & Osuzuka (1977)
2	400	–	Rathus & Landy (1961)
2	250	–	Hine (1969)
2	176	–	Rathus & Landy (1961)
3	400	–	Rathus & Landy (1961)
4	120	–	Rathus & Landy (1961)
4	72	–	Rathus & Landy (1961)
4	60	–	Rathus & Landy (1961)
5	120	–	Rathus & Landy (1961)
6	250	–	Rathus & Landy (1961)
6	–	300	Viner (1945)
9	48	–	Rathus & Landy (1961)
13	84	–	Rathus & Landy (1961)
13	48	–	Rathus & Landy (1961)
38	0	–	Rathus & Landy (1961)
43	15		Rathus & Landy (1961)
46	10	–	Rathus & Landy (1961)
81	0	–	Rathus & Landy (1961)

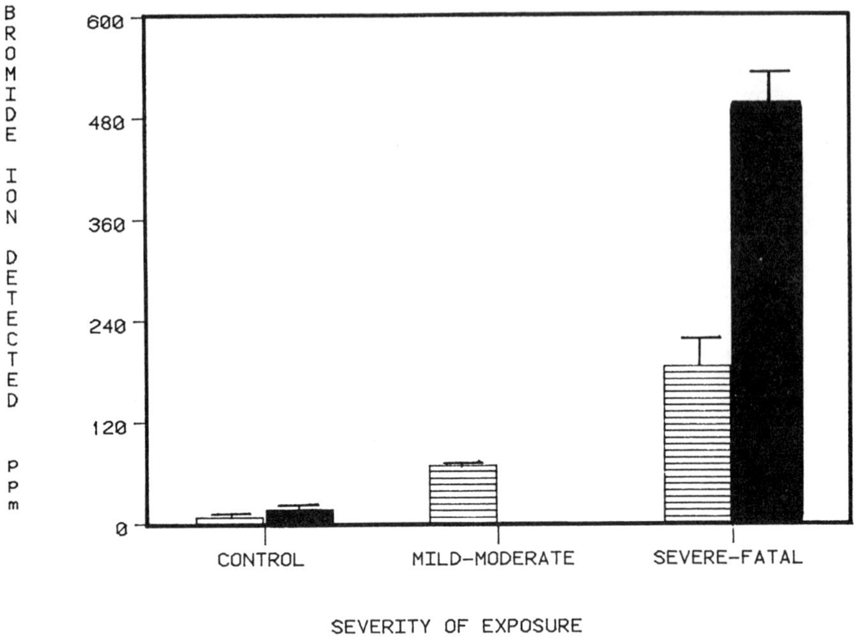

Fig. 1. Human bromide ion concentration determined following acute methyl bromide exposure. The values used were those reported in the literature within three days following methyl bromide exposure, and presented in Table IV. The mean bromide ion values (± SE) are plotted for unexposed controls, and acute cases where the symptoms reported were either mild to moderate or severe to fatal. The blood bromide ion means were compared using a one-way analysis of variance and were significantly different; p = 0.02. Note: Bromide ion analysis in the serum of patients with mild to moderate symptoms have not been reported. ▤ = blood bromide, ■ = serum bromide.

ion determinations (± SE) following acute exposure are presented in Figure 1. The mean blood bromide ion level for controls was 6.7 ppm, while for patients with mild to moderate symptoms it was 69 ppm, and for those with severe to fatal exposures it was 157 ppm. The means were compared using analysis of variance and the differences were statistically significant at the p=0.02 level. Multiple comparison procedures indicate that the major difference is between the control and the severe to fatal exposures cases. Bromide ion values are not available in serum for the patients with mild to moderate symptoms following an acute exposure, since none of the literature values were measured within three days of exposure. Consequently the acute serum bromide ion levels were not analyzed statistically.

The pooled bromide ion levels in blood and serum following chronic exposure are presented in Figure 2. Persons chronically exposed to methyl bromide with no symptoms had a mean blood bromide ion value of 10.7 ppm (Rathus and

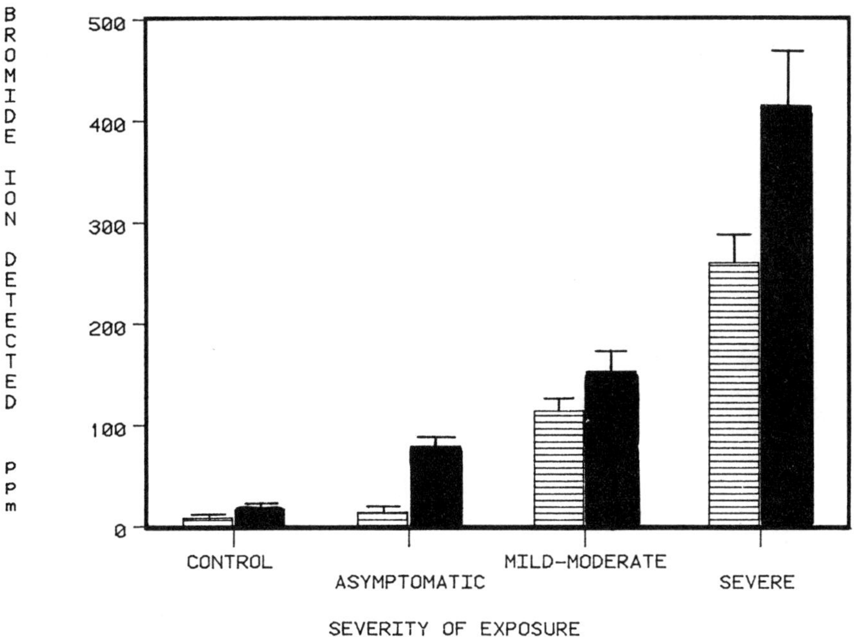

Fig. 2. Human bromide ion concentrations determined following chronic methyl bromide exposure. The values used were those reported in the literature within three days following methyl bromide exposure. The mean bromide ion values (± SE) are plotted for unexposed controls, and persons who were chronically exposed and exhibited no symptoms, mild to moderated symptoms, or severe to fatal symptoms. The blood and serum bromide ion means were compared using a one-way analysis of variance, and were significantly different; $p < 0.001$. ▤ = blood bromide, ■ = serum bromide.

Landy 1961, Verberk *et al.* 1979) and a mean serum bromide ion level of 80 ppm (Drawneek *et al.* 1964). Persons chronically exposed to methyl bromide with mild to moderate symptoms had a mean blood bromide level of 114 ppm and a mean serum bromide ion level of 154 ppm. For severe and fatal chronic exposure cases, the mean bromide ion levels were 260 ppm in blood and 414 ppm in serum.

When the bromide ion concentrations in either blood or serum are compared using the analysis of variance, they are very significantly higher than controls at the $p < 0.001$ level. A multiple comparison procedure indicates that when controls are compared with those persons who had mild to moderate symptoms, the exposed persons have bromide ion levels significantly greater than those of controls at the $p < 0.02$ level. When bromide ion levels of those who had severe or fatal symptoms are compared with controls, they are significantly higher at the $p < 0.001$ level.

In persons with no symptoms, the bromide ion levels were elevated, but not statistically different from controls. In general, exposure to methyl bromide is associated with increased bromide ion levels. Therefore the determination of bromide ion levels may be useful in ascertaining the severity of exposure, particularly in chronic cases.

Following an acute methyl bromide exposure, the bromide ion level may be elevated for more than 35 days. Clarke *et al.* (1945) reported that the blood bromide ion levels returned to normal following 32 to 40 days. Benatt and Courtney (1948) reported that the serum bromide ion level was still substantially elevated at 19 days post-exposure. In contrast, the blood bromide ion levels may not return to normal for 60 days in cases of chronic methyl bromide exposure (Rathus and Landy 1961).

As discussed above, it is possible that the chronically exposed actually develop a low-grade lesion. A study by Verberk *et al.* (1979) reported that fumigators chronically exposed to methyl bromide, although not exhibiting any obvious symptoms or pathology, had elevated bromide ion levels correlated with slight electroencephalographic changes and a small increase in serum transaminases. It seems clear that those persons who are exposed chronically to methyl bromide (fumigators) should periodically have their serum or blood bromide ion levels measured to determine if it is substantially greater than control levels, particularly if they are exhibiting any abnormal symptoms. A monitoring program has been undertaken in Queensland, Australia (Rathus 1980). Persons with blood bromide ion levels greater than 20 ppm are removed from exposure. The value was adopted on the basis of evidence of toxicity at blood bromide levels of 28 ppm (Rathus and Landy 1961).

Therapy for methyl bromide poisoning has basically been supportive, particularly with attempts to control the seizures and convulsions (Miller 1943, Holling and Clarke 1944, Carter 1945, Clarke *et al.* 1945, Wyers 1945, MacDonald *et al.* 1950, von Oettingen 1958, Longley and Jones 1965, Hine 1969, Greenberg 1971, Goulon *et al.* 1975), but some experimental therapies have been conducted. Rathus and Landy (1961) treated their seven patients with BAL, an antidote for arsenic poisoning. Since all patients survived, the physicians felt that the treatment was beneficial, possibly preventing the development of irreversible symptoms. Shield *et al.* (1971) used BAL in treatment of one case where the patient eventually improved. Zatuchni and Hong (1981) used acetylcysteine in treatment of methyl bromide poisoning and concluded that the treatment was not harmful and probably beneficial. Bezuglyi *et al.* (1977) reported that administration of cysteine to two patients acutely exposed to methyl bromide dramatically improved their recovery. In both cases the patients suffered from moderate acute poisoning. Following four days of treatment with cysteine the patients became asymptomatic and did not have any recurring symptoms. Obviously, further investigations of the antidotal treatment of cysteine, acetylcysteine, and BAL for methyl bromide poisoning is necessary.

XVIII. Residues in food

The potential exposure of the consumer to methyl bromide in foodstuffs has been investigated. Laug (1941) reported that 85 to 95% of the volatile methyl bromide residue disappeared from the foodstuffs within one hr. After 24 hr, the residue was reduced even further. Walnuts and cheese retained the highest methyl bromide residue concentrations, particularly bromide ion. The author concluded that the higher residues are due to these foods' relatively high fat content, but in retrospect it may have been due to their higher protein content. Shrader *et al.* (1942) showed that the methyl bromide residues decreased to control levels in flour and processed cheese after seven days. Page *et al.* (1949) determined the volatile residues in dried fruit following fumigation. In raisins the methyl bromide residue was less than four ppm immediately after fumigation (2 lb/1,000 cu ft) and after two days the level was below one ppm. A report by Lindgren *et al.* (1962) indicated that most of the volatile bromide had dissipated within six hr following fumigation of wheat (2 lb/1,000 cu ft). Rhodes *et al.* (1975) found that after one mon, 2.3 ppm of methyl bromide were detected in fumigated frozen uncooked beef fat, while 0.4 ppm was detected after three mon.

A study by Scudamore and Heuser (1970) considered the methyl bromide residue in a variety of commodities with different moisture contents and at different temperatures. Generally methyl bromide dissipated more quickly at higher temperatures and in commodities with higher moisture content. Groundnuts and cottonseed cake retained substantially higher residue levels than wheat flour, Manitoba wheat, yellow maize, sorghum, sultanas, and cocoa beans.

In an effort to determine if methyl bromide residues from fumigation were harmful, food fumigated with methyl bromide (3 lb/1,000 cu ft) was fed to rats and rabbits for 20 and 52 wk, respectively. No appreciable toxicologic effects were observed but the exposed rabbits had a slightly lower weight gain than the controls. Authors have generally agreed that the low residue levels of methyl bromide provide little risk to the consumer (Dudley and Neal 1942, Page *et al.* 1949, Scudamore and Heuser 1970). Methods for determining total bromide and organic bromide in foods were recently reviewed by Getzendaner (1975).

Summary of current information

Over the past hundred yr of methyl bromide use, information on its toxic effects has primarily been obtained from medical reports on human poisoning cases. Methyl bromide has produced over 950 documented poisoning incidents in human beings. These cases have included fatalities, systemic poisonings, skin injuries, and eye injuries. Fewer than ten fatalities have been reported in the last ten yr. The route of methyl bromide exposure is primarily by inhalation, but dermal exposure is also possible. The primary health hazard is acute or chronic exposure to workers during fumigation. The concentration necessary to produce an onset of acute symptoms is poorly defined. Most studies of acute animal exposures have reported concentrations necessary to produce 100% mortality, but few LC_{50}s for specific exposure conditions have been reported.

Methyl bromide is neurotoxic and a variety of symptoms have been reported in human beings, including headache, dizziness, fainting, speech impairment, mood swings, loss of memory, mental confusion, vision impairment, hearing disorders, vertigo, giddiness, delirium, stupor, apathy, lethargy, limb weakness, limb numbness, pain in the extremities, tremors, muscle twitching, paralysis, convulsions, seizures, ataxia, incoordination, narcosis, and unconsciousness. Careful examination shows that certain neurological symptoms can be attributed to either acute or chronic exposure, but persons acutely exposed have on occasion developed "chronic" symptoms.

Animal experimentation has in general confirmed the ability of methyl bromide to produce severe acute and chronic neurological signs such as tremors, paralysis, ataxia, and convulsions. Only recently have experiments begun to consider effects on more sensitive measures such as behavior.

Non-neurological symptoms have been reported in human beings, particularly effects on the respiratory tract, and occasionally effects on the kidney. In animals, except for the respiratory system the non-neurological effects of methyl bromide have generally been ignored. A delayed onset of toxicity has on occasion been noted, particularly at lower exposure concentrations.

The alkylating ability of methyl bromide has been demonstrated in a variety of *in vitro* experiments. This property is most likely associated with methyl bromide's mechanism of toxic action, but the compound's lack of specificity suggests that many biological changes may be occurring simultaneously. Methyl bromide has been shown to be mutagenic and it should be studied for carcinogenicity.

References

Anderson, M. E., M. L. Gargas, R. A. Jones, and L. J. Jenkins, Jr.: Determination of the kinetic constants for metabolism of inhaled toxicants *in vivo* using gas uptake measurements. Toxicol. Applied Pharmacol. **54**(1), 100 (1980).

Andrieu, G., P. Meriel, and Bourbon: Deux cas d'intoxication par le bromure de methyle. Traitment par reanimation respiratoire. Arch. Mal. Prof. **20**, 776 (1959).

Anger, W. K., J. V. Setzer, J. M. Russo, W. S. Brightwell, R. G. Wait, and B. L. Johnson: Neurobehavioral effects of methyl bromide inhalation exposures. Scand. J. Work Environ. Health. **7** (Suppl. 4), 40 (1981).

Anonymous: Fumes on the Peninsula–18 in hospital. *San Franciso Chronicle.* Oct. 6, p. 1 (1973).

Araki, S., K. Ushio, K. Suwa, A. Abe, and K. Uehara: Methyl bromide poisoning: A report based on fourteen cases. Sangyo Igaku. **13**(6), 507 (1971).

Auck, S. E.: Short history of halogenated fire extinguishing agents. *An Appraisal of Halogenated Fire Extinguishing Agents.* Washington: Nat. Acad. Sci. (1972).

Bachem, C.: Contribution to the toxicology of halogen alkyls. Arch. Exp. Pathol. Pharmakol. **122**, 69 (1927) [cited in von Oettingen (1946)].

Bakhishev, G. N.: Effect of cysteine on the level of some functional groups of proteins during acute poisoning by methyl bromide. Gig. Primen. Toksikol. Pestits. Klin. Otravl. **9**, 261 (1971) [cited in Chem. Abstr. **77**, 148215J (1972).

______ Relative toxicity of aliphatic halohydrocarbons to rats. Farmakol. Toksikol. (Kiev). **8**, 140 (1973).

______ Cysteine activity in animal intoxication with different aliphatic series haloid-hydrocarbons. Farmakol. Toksikol. (Moscow). **41**(3), 342 (1978).

Baynes-Cope, A.: The choice of biocides for library and archival material. Biodeterior. Mater. Proc. Int. Biodeterior. Symp. 2nd, 1971, p. 381 (1972).

Benatt, A. J., and T. R. Courtney: Uraemia in methyl bromide poisoning: A case report. Brit. J. Ind. Med. **5**(21), 21 (1948).

Berry, C. E.: Methyl bromide as a rodenticide. Calif. Dept. Agr. Bull. **27**(1), 173 (1938).

Bernot, A., and B. Goldberg: Intoxication par le bromure de methyle. Bull. Soc. Med. Milit. France. **54**, 221 (1960).

Beumie, L.: Intoxication par le bromure de methyle en milieu militaire. Soc. Med. de Travail de Toulouse, 27 juin 1953. Compte Rendu de Press Med. **61**(68), 1392 (1953).

Beyne and Goett: Toxicity of carbon tetrachloride and methyl bromide used in fire extinguishers, and precautions for their use. Arch. Med. Pharm. Naval. **124**, 409 (1934) [cited in von Oettingen (1946)].

Bezuglyi, V. P., I. G. Mizyukova, G. N. Bakhishev, and E. A. Afanas'eva: Use of cysteine in the treatment of acute methyl bromide poisoning. Vrach. Delo. **4**, 124 (1977).

Billet and Abel: Accidents consecutifs a l'emploi d'un extincteur d'incendie charge au bromure de methyle. Gaz. Hop. Civ. Mil. **113**(45-46), 464 (1940).

Blackburn, S., E. Carter, G. Hemphill, and H. Phillips: The methylation of wool with methyl sulphate and methyl halides. Biochem. J. **35**, 627 (1941).

______, and H. Phillips: Experiments on the methylation and acetylation of wool, silk fibroin, collagen and gelatin. Biochem. J. **38**, 171 (1944).

______, R. Consden, and H. Phillips: The action of sulphites on cysteine disulphide linkages of wool. 4. Methylation of the thiol groups of bisulphited wools. Biochem. J. **38**, 25 (1944).

Bond, E. J., and C. T. Buckland: Control of insects with fumigants at low temperatures: Toxicity of fumigants in atmospheres of carbon dioxide. J. Econ. Entomol. **71**(2), 307 (1978).

Botteri, B. P., R. E. Cretcher, and W. R. Kane: Aircraft applications of halogenated hydrocarbon fire extinguishing agents. *An Appraisal of Halogenated Fire Extinguishing Agents*. Washington: Nat. Acad. Sci. (1972).

Bowley, C. R., and C. H. Bell: The toxicity of twelve fumigants to three species of mites infesting grain. J. Stored Prod. Res. **17**(2), 83 (1981).

Bridges, R. G.: The fate of labelled insecticide residues in food products. III. N-Methylation as a result of fumigating wheat with methyl bromide. J. Sci. Food Agr. **6**, 261 (1955).

Butler, E. C. B., K. M. A. Perry, and J. R. F. Williams: Methyl bromide burns. Brit. J. Ind. Med. **2**, 30 (1945).

Cade, A., and P. Mazel: Intoxication par le bromure de methyle. Soc. Med. Hop. Paris, Bull. Mem. **47**, 722 (1923).

California Department of Food and Agriculture: Pesticide use report. Annual (1970 through 1981).

______ Agriculture chemicals and feed. Safety precautions on the use of methyl bromide. Pesticide Safety Information Series No. 5 (1974).

California Department of Public Health, Bureau of Occupational Health and Environmental Epidemiology: Occupational disease in California attributed to pesticides and other agricultural chemicals (1957 through 1970 and 1971-1973).

Carter, A. B.: Methyl bromide poisoning: Effects on the nervous system. Brit. Med. J. **1**, 43 (1945).

Castaing, R., J. Cardinaud, J. Favarel-Garrigues, P. Chomy, G. Gillardeau, and M. Bougeois: L'intoxication aigue au bromure de methyle. J. Med. Bordeaux. **143**, 1757 (1966).

Chandler, J. M.: Methyl bromide as a refrigerating agent. U.S. Pat. No. 1,547,202, July 2, 1925 [cited in Chem. Abstr. **19**, 2858 (1925)].

Clarke, C. A., C. G. Roworth, and H. E. Hollling: Methyl bromide poisoning. Brit. J. Ind. Med. **2**, 17 (1945).

Collins, R. P.: Methyl bromide poisoning, a bizarre neurological disorder. Calif. Med. **103**, 112 (1965).

Conway, E. J., and J. C. Flood: An absorption apparatus for the micro-determination of certain volatile substances. The micro-determination of bromide, with application to blood and urine and observations on the normal human subject. Biochem. J. **30**, 716 (1936).

Corcos, A., J. Heuretematte, and V. Corcos: Intoxication mortelle par le bromure de methyle. Bull. Mem. Soc. Med. Hop. Paris. **71**, 1005 (1955).

Davay, G. G.: Methyl bromide poisoning. Indian J. Ind. Med., **18**(2), 78 (1972) [cited in Pest. Abstr. **74**, 832 (1972)].

Dawson, G. W., A. L. Jennings, D. Drozdowski, and E. Rider: The acute toxicity of 47 industrial chemicals to fresh and saltwater fishes. J. Hazard. Mater. **1**(4), 303 (1977).

de Gaulejac, R., and P. Dervillee: Les manifistations toxiques, generales et locules dues au bromure de methyle au cours de son emploi comme extincteur d'incendie. Gaz. Hebdomadaire Sciences Med. Bourdeaux. **10**, 149 (1939).

Dejong, R. N.: Methyl bromide poisoning with special reference to nervous system manifestations. J. Amer. Med. Assoc. **125**(10), 702 (1944).

de Morsier, G., and J. Steinmann: Epilepsie chronique consecutive a l'intoxication par le bromure de methyle. Revue des accidents du travail et des maladies professionnelles, Berne, 1936. Ann. Hyg. Publique Ind. Soc. **14**, 718 (1936).

Derobert, L.: Brulures cutanees consecutives a l'emploi du bromure de methyle comme extincteur d'indendie. Arch. Mal. Prof. Med. Trav. Secur. Soc. **3**, 54 (1941).

______, R. Martin, and R. Lebreton: A propos d'une intoxication par le bromure de methyle. La Presse Medicale. **61**(39), 819 (1953).

Dieter, C. E., and L. L. Coulter: Dowfume MC-2 for fumigation of plant beds. Down to Earth **4**, 2 (1949).

Dixon, M., and D. M. Needham: Biochemical research on chemical warfare agents. Nature **158**, 432 (1946).

Dixon, T. F.: Bromine in the tissues. Biochem. J. **29**, 86 (1935).

Djalali-Behzad, G., S. Hussain, S. Osterman-Golkar, and D. Segerbaeck: Estimation of genetic risks of alkylating agents. VI. Exposure of mice and bacteria to methyl bromide. Mutat. Res. **84**(1), 1 (1981).

Drawneek, W., M. J. O'Brien, H. J. Goldsmith, and R. E. Bourdillon: Industrial methyl bromide poisoning in fumigators. Lancet **2**, 855 (1964).

Dubrisay, J., and J. Fages: La pathologie professionnelle dans les activites agricules. Essai d'approche statistique. II. Les maladies d'origine professionnelle. Arch. Mal. Prof. Med. Trav. Secur. Soc. **39**(7-8), 459 (1978).

Dudley, H. C., and P. Neal: Methyl bromide as a fumigant for food. J. Food Sci. **7**(6), 421 (1942).

Dunkelberg, H.: Zur Problematik per Anwendung alkylierender Substanzen bei der Lebensmittelbegasung hinsichtlich der Bildung von Vorstufen der N-Nitrosoverbindungen. II. N-Methylierung verschiedner Aminosauren durch Einwirkung von Methylbromid. Zbl. Bakt. Hyg. I. Abt. Orig. **171**, 48 (1980).

Duvoir, M., R. Fabre, and F. Layani: L'intoxication par le bromure de methy. Bull. Mem. Soc. Med. Hop. Paris. **34**, 1540 (1937).

——— ——— L'intoxication par le bromure de methyle. Bull. Sci. Pharmacol. **46**, 15 (1939).

Eckardt, R. E.: Industrial intoxications which may simulate ethyl alcohol intake. Ind. Med. Surg. **40**(3), 33 (1971).

Ehrenberg, L., S. Osterman-Golkar, D. Singh, and U. Lundquist: Reaction kinetics and mutagenic activity of methylating and beta-halogenoethylating gasoline additives. Radiat. Bot. **14**, 185 (1974).

Fisk, F., and H. Shepard: Laboratory studies of methyl bromide as an insect fumigant. J. Econ. Entomol. **31**(1), 79 (1938).

Florentin, D.: Le bromure de methyle: Un noveau toxique industriel insidiex. Ann. Hyg. Publique Ind. Soc. **22**, 142 (1944).

Ford, C. L.: An overview of halon 1301 systems. In R. G. Gann (ed.): Halogenated fire suppressants **16**, pp. 1-8. Amer. Chem. Soc. Symp. Series (1975).

Franken, L., P. Bastenie, and L. Van Bogaert: Etude anatomique d'un cas d'intoxication par le bromure de methyle. Acta Neurol. Psych. Belg. **59**, 375 (1959).

Friedman, M., and M. R. Gumbmann: Biological availability of epsilon-N-methyl-L-lysine, 1-N-methyl-L-histidine, and 3-N-methyl-L-histidine in mice. Nutr. Rep. Int. **19**(4), 437 (1959).

Gallo, F. P.: Bromuro di metile ossido di etilene formaldeide: problemi biologici, tossicologici e problemi correlati al trattemento dei materiali librari. Parte prima. Nuovi Ann. Ig. Microbiol. **29**(1), 51 (1978 a).

——— Parte Seconda. Nuovi Ann. Ig. Microbiol. **29**(2), 181 (1978 b).

Garcia Rico, A. M., J. Garces Bruses, J. Mas Marfany, and R. Nolla Pandes: Intoxicacion aguola collectiva por bromura de metilo. Med. Clin. (Barcelona) **63**(6), 291 (1974).

Gargas, M. L., M. E. Andersen, and L. J. Jenkins, Jr.: Metabolism of inhaled organic bromides. Corroboration of gas uptake results by direct measurement of inorganic bromide production. Soc. Toxicol., 19th Ann. Meeting, Abstr. 118 (1980).

Gaultier, M., E. Fournier, P. Gervais, M. L. Efthymiou, and J. P. Frejauille: 205 intoxications aigues accidentelles par les produits industriels, observees en reanimation a l'hopital Fernand-Widal de 1964 a 1973. J. Eur. Toxicol. **6**(6), 263 (1973).

Gayral, M.: Troubles vaso-moteurs graves de l'estremite cephalique par intoxication par bromure de methyle. Arch. Mal. Prof. Hyg. Toxicol. Ind. **10**, 358 (1949).

Getzendaner, M. E.: Bromide residues in chicken tissues and eggs from ingestion of methyl bromide-fumigated feed. J. Agr. Food Chem. **13**(4), 349 (1965).

______ A review of bromine determination in foods. J. Assoc. Official Anal. Chemists **58**(4), 711 (1975).

Girard, P. F., M. Tommasi, L. Rodre, and E. Lejeune: Documents anatomiques concernant les lesions de systeme nerveux central dans un cas d'intoxication aigue par le bromure de methyle. Rev. Neurol. **103**, 165 (1960).

Glaser, E., and S. Frisch: Ein beitrag zur kenntnis der wirkung technisch und hygienisch wichtiger gase und dampfe auf den organismus. Uber gebromte kohlenwasserstoffe de fettreihe. Arch. Hyg. Bakteriol. **101**, 48 (1929).

Goulon, M., F. Nouailhat, R. Escourolle, J. J. Zarranz-Imirizaldu, S. Grosbuis, and M. A. Levy-Alcover: Intoxication par le bromure de methyle. Trois observations, dont use mortelle. Etude neuropathologique d' un cas de stupeur avec myloclonies, suivi pendant cinq ans. Rev. Neurol. **131**(7), 445 (1975).

Gray, P. H. K.: Methyl bromide poisoning. J. R. Nav. Med. Serv. **30**, 214 (1944).

Greenberg, J. O.: The neurological effects of methyl bromide poisoning. Ind. Med. Surg. **40**, 27 (1971).

Grossetette, G., J. Larribaud, G. Chevrel, J. Hainaut, P. Colonna, M. Le Clainche, C. Brucher, R. Lefebvre, and Y. Lefebvre: Les formes graves de l'intoxication au bromure de methyle. A propos de cinq observations. Presse Med. 2, 1870 (1963).

Hague, N. G., and U. Sood: Soil sterilization with methyl bromide to control soil nematodes. Plant Pathol. **12**, 88 (1963).

Hammon, J., J. Paraire, J. Velluz, and L. Damasio: L'intoxication aigue par le bromure de methyle remarques semeilogigues et etude electroencephalographique. A propos d'un cas. Sem. Hop. Paris. **31**, 907 (1955).

Hardin, B. D., G. P. Bond, M. R. Sikon, F. D. Andrew, R. P. Beliles, and R. W. Neimeier: Testing of selected workplace chemicals for teratogenic potential. Scand. J. Work Environ. Health **7** (Suppl. 4), 66 (1981).

Harry, E. G., and W. B. Brown: Fumigation with methyl bromide: Applications in the poultry industry: A review. World's Poult. Sci. J. **30**(3), 193 (1974).

Heise, H., and A. Moller: Bericth über eine dermale Intoxikation mit Methylbromid. Dermatol. Monatsschr. **162**(10), 837 (1976).

Hemming, A.: An uninflammable volatile liquid consists of a mixture of MeCl and MeBr. Canadian Pat. 213,825, Oct. 18, 1921 [reported in Chem. Abstr. **18**, 1301 (1921)].

Henning, A.: Application of the halogen derivatives of the hydrocarbons with particular reference to methyl bromide. Chem. Ind. London **52**(22), 462 (1933).

Hine, C. H.: Methyl bromide poisoning: A review of ten cases. J. Occup. Med. **11**(1), 1 (1969).

Holling, H. E., and C. A. Clarke: Methyl bromide intoxication. J. R. Nav. Med. Serv. **30**, 218 (1944).

Homma, T., A. Sudo, M. Miyagawa, M. Sato, H. Hasegawa, and M. Suda: Studies on intoxication due to methyl bromide (3)–Effects of methyl bromide on the central nervous system. Sangyo Igaku Sogo-Kenkyusho Nempo., p. 22 (1980).

Hunter, D.: The diseases of occupations, p. 561. London: Hodder and Stoughton (1975).

Ikeda, T., R. Kishi, K. Yamamura, H. Miyake, M. Sato, and S. Ishizu: Behavioral effects in rats following repeated exposure to methyl bromide. Toxicol. Lett. (Amst.) **6**(4-5), 293 (1980).

Ingram, F. R.: Methyl bromide fumigation and control in the date-packing industry. Ind. Hyg. Occup. Med. **4**(3), 193 (1951).

Irish, D. D., and E. M. Adams: Apparatus and methods for testing toxicity of vapors. Ind. Med. **9** (Ind. Hyg. Sec), 1 (1940).

______ ______, H. C. Spencer, and V. K. Rowe: Response attending exposure of laboratory animals to vapors of methyl bromide. J. Ind. Hyg. Toxicol. **22**, 218 (1940).

______ ______ ______ ______ Chemical changes of methyl bromide in the animal body in relation to its physiological effects. J. Ind. Hyg. Toxicol. **23**(8), 408 (1941).

Irsigler, F. J.: A case of methyl bromide poisoning simulating rupture of an intracranial aneurysm. S. Afr. Med. J. **25**(47), 949 (1951).

Ishizu, S., H. Momotani, M. Sato, and H. Uchiyama: On the acute intoxication due to methyl bromide. Sangyo Igaku **21**(4), 662 [cited in Pest. Abstr. **80**, 848 (1979)].

Jameson, H. D.: EEG signs in methyl bromide intoxication. Electroencephalogr. Clin. Neurophysiol. **46**(1), 2P (1979).

Jaquet, A.: Ueber Brommethylvergifung. Deutsches Archiv. Klinische Medizin **71**, 370 (1901).

Joachimoglu, G.: Toxikologische Betrachtungen ueber einige modern Feuerloschmittel. Dtsch. Med. Wochenschr. **56**, 785 (1930).

Johnstone, R. T.: Methyl bromide intoxication. Ind. Med. **14**(6), 495 (1945).

Jones, G. W.: The flammability of refrigerants: mixtures of methyl and ethyl chlorides and bromides. Ind. Eng. Chem. **20**(4), 367 (1928).

Jordi, A. U.: Absorption of methyl bromide through the intact skin: a report of one fatal and two non-fatal cases. Aviat. Med. **24**, 536 (1953).

Kashima, T., M. Fukui, C. Wakasugi, K. Nishimoto, H. Yamano, and K. Fukuda: A fatal case of acute methyl bromide poisoning. Nippon Hoigaku Zassh. **23**, 241 (1969) [cited in Chem. Abstr. **72**, 88438u (1969)].

Kawai, H., and K. Ueda: Effect of glutathione on acute intoxication due to methyl bromide. Nippon Noson Igakkai Zasshi. **21**(2), 314 [cited in Pest. Abst. **73**, 399 (1972)].

Kantarjian, A. D., and A. S. Shaheen: Methyl bromide poisoning with nervous system manifestations resembling polyneuropathy. Neurology **13**, 1054 (1963).

Knight, H. D., and M. Reina-Guerra: Intoxication of cattle with sodium bromide-contaminated feed. Amer. J. Vet. Res. **38**(3), 407 (1977).

Langlois, M., R. M. Vercel, R. Kahlil, and Bereni: Sur quatre cas d'intoxication collective par le bromure de methyle. Considerations cliniques et electroencephalographiques. Rev. Neurol. **108**, 305 (1963).

Laug, E. P.: Bromide residues in foodstuffs. Volatile and nonvolatile residues following experimental exposure to methyl bromide. Ind. Eng. Chem. **33**, 803 (1941).

Lawley, P. D.: Carcinogenesis by alkylating agents. In D. Searle (ed.): Chemical carcinogens, pp. 83-244. Washington: Amer. Chem. Soc. (1976).

Le Goupil, M.: Les proprieties insecticide du bromure de methyle. Rev. Pathol. Veg. Entomol. Agr. Fr. **19**, 169 (1932).

Lewis, S. E.: Inhibition of SH enzymes by methyl bromide. Nature **161**, 692 (1948).

______, and K. Eccleston: Residues in wheat flour fumigated with methyl bromide. J. Soc. Chem. Ind. **65**, 149 (1946).

Lindgren, D. L., Gunther, F. A., and L. E. Vincent: Bromide residues in wheat and milled wheat fractions fumigated with methyl bromide. J. Econ. Entomol. **55**, 773 (1962).

Longley, E. O., and A. T. Jones: Methyl bromide poisoning in man. Ind. Med. Surg. **34**, 499 (1965).

______, and G. R. Simpson: Acute methyl bromide poisoning. Proc. Int. Congress Rural Med., 4th, p. 32 (1970).

Maag, T. A., and S. C. Schmittle: The effect of methyl bromide upon *Salmonella pullorum*. Amer. J. Vet. Res. **23**, 1289 (1962).

MacDonald, A. C., I. C. Monro, and G. I. Scott: Fatal case of poisoning due to inhalation of methyl bromide. Brit. Med. J. **2**, 441 (1950).

Mackie, D. B.: Methyl bromide: Its expectancy as a fumigant. J. Econ. Entomol. **31**(1), 70 (1938).

______, and W. B. Carter: Methyl bromide as a fumigant: A preliminary report. Calif. Dept. Agr. Bull. **26**(2), 153 (1937).

Maddy, K. T.: Pesticides reported by physicians in California as causes of illness in employed persons in 1974. Pesticide Illness Report. No. HS-171, p. 4. Sacramento: Calif. Dept. Food Agr. (1976).

______, S. A. Peoples, and D. Berry: Pesticide-related human illnesses in California, Vol. I (1975), Vol. II (1976), Vol. IV (1978), Vol. V (1979), Vol. VI (1980). *Calif. Dept. Food Agr., Worker Health and Safety Unit.*

Marchand-Aliphant, A and M.: Un cas d'intoxication mortelle par le bromure de methyl d'un extincteur. Ann. Med. Leg. **32**, 162 (1952).

Mazel, P., J. Bourret, and L. Roche: Intoxication familiale de cinq personnes au cours d'une desinsectisation par le bromure de methyl. Trois morts. Arch. Mal. Prof. **7**, 38 (1946).

Mellerio, F., M. Bautier, and C. Bismut: Electroencephalographie au cours des intoxications aigues par bromure de methyle. J. Eur. Toxicol. **7**(2), 119 (1974).

Menge, J. A., D. E. Munnecke, E. L. Johnson, and D. W. Carnes: Dosage response of the vesicular-arbuscular mycorrhizal fungi *Glomus fasciculatus* and *G. constrictus* to methyl bromide. Phytopathol. **68**(9), 1368 (1978).

Merzback, L.: Zur pharmakologie des Brommethyls and einiger seiner Verwandten. Z. Exp. Med. **63**, 383 (1928).

Miller, B. H., R. Navone, and M. Ota: Irritation from residual bromides after methyl bromide fumigation. Public Health Rep. **76**, 216 (1961).

Miller, D. P., and H. W. Haggard: Intracellular penetration of bromide as a feature in the toxicity of alkyl bromides. J. Ind. Hyg. Toxicol. **25**(10), 423 (1943).

Miller, J. W.: Fatal methyl bromide poisoning. Arch. Pathol. **36**, 505 (1943).

Miyagawa, M., H. Hasegawa, M. Sato, A. Sudo, T. Homma, and K. Okonogi: Studies on intoxications due to methyl bromide. Proc. Ann. Meet. Jpn. Soc. Ind. Hyg. **53**, 309 (1980)[cited in Pest. Abst. **80**, 2336 (1980)].

________ Conditioned taste aversion induced by inhalation exposure to methyl bromide in rats. Toxicol. Letters **10**, 411 (1982).

Mosinger, M., H. Ollivier, H. Fiorentini, M. Rossano, and F. Vuillet: Sur la pathologie due au bromure de methyle. A propos de trois observations. Ann. Med. Leg. **4**, 257 (1964).

Munnecke, D. E., J. L. Bricker, and M. J. Kolbezen: Comparative toxicity of gaseous methyl bromide to ten soilborne phytopathogenic fungi. Phytopathol. **68**(8), 1210 (1978).

Nishimura, M., M. Umeda, S. Ishizu, and M. Sato: Cytotoxicity of methyl bromide: Effect of methyl bromide on cultured mammalian cells. J. Toxicol. Sci. **5**(4), 321 (1980).

Nuckolls, A. H.: Report on the comparative life, fire, and explosion hazards of common refrigerants. Underwriters' Laboratories Report on Miscellaneous Hazards No. 2375 (1933).

Page, A. B. P., O. F. Lubatti, and J. Russell: Application of fumigants to ships and warehouses: VI. Fumigation of dried fruit with methyl bromide in chambers. J. Soc. Chem. Ind. **68**, 102 (1949).

Pavkov, K. L., W. D. Kerns, C. E. Chrisp, D. C. Thake, R. L. Persing, and H. H. Harroff: Major findings in twenty-four month inhalation toxicity study of methyl chloride in mice and rats. Toxicologist **2**, 161 (1982).

Pernod, J., M. Tommasi, R. Damasio, and J. Magerand: Intoxication aigue collective par le bromure de methyle. Bull. Soc. Med. Paris. **7**, 235 (1961).

Powell, D. F.: The effects on narcissus bulbs of methyl bromide fumigation used to control bulb scale mite. Plant Pathol. **26**(2), 79 (1977).

Prain, J. H., and G. H. Smith: Methyl bromide: Clinicopathologic report of eight cases of poisoning. Brit. J. Ind. Med. **9**, 44 (1952).

Rathus, E. M.: Pesticide exposure in tropical and sub-tropical Australia. In W. F. Tordoir and E. Van Heemstra (eds.): Field workers exposure during pesticide application, pp. 85-91. New York: Elsevier (1980).

_______, and P. J. Landry: Methyl bromide poisoning. Brit. J. Ind. Med. **18**, 53 (1961).

Rauws, A. G.: Bromide pharmacokinetics: A model for residue accumulation in animals. Toxicol. **4**, 195 (1975).

_______, and M. J. Van Logten: The influence of dietary chloride on bromide excretion in the rat. Toxicol. **3**, 29 (1975).

Rhodes, D. N., R. L. S. Patterson, D. J. Puckey, S. G. Heuser, H. E. Wainman, B. Chakrabarti, and E. N. W. Allan: Exposure of meat to methyl bromide during refrigerated-store pest control; residues in beef and lamb and effect on quality. J. Sci. Food Agr. **26**, 1375 (1975).

Richardson, B. W.: Some further additions to therapeutics. Practitioner **6**, 337 (1871).

______ Methyl bromide. Asclepiad (London) **8**, 239 (1891).

Roche, L., M. Colin, M. Tommasi, E. Legeune Moitrepierre, and Grandmottet: Intoxication mortelle par le bromure de methyle. Manifestations renales, pulmonaires, neurologiques. Prolongation du coma a la suite de la therapentique. Ann. Med. Leg. **38**, 364 (1958).

Roger, H., and E. Hawthorn: L'intoxication par le bromure de methyle. Presse Med. **40**, 1178 (1941).

Roycroft, J. H., R. H. Jaskot, E. C. Grose, and D. E. Gardener: The effects of inhalation exposure of methyl bromide in the rat. Toxicologist **1**, 79 (1981).

Sato, M., H. Hasegawa, T. Homma, M. Miyagawa, M. Suda, and A. Sudo: Studies on intoxication due to methyl bromide (2)–distribution of methyl bromide and bromine in the living body. Sangyo Igaku Sogo-Kenkyusho Nempo, p. 22 (1980 a)[cited in Pest. Abstr. **81**, 3225 (1980)].

______, H. Uchiyama, and S. Marinobu: Experimental studies of intoxication due to methyl bromide centered on the persistence of bromine in organs. Proc. Ann. Meet. Jpn. Soc. Ind. Hyg. **53**, 305 (1980 b) [cited in Pest. Abst. **80**, 2334 (1980)].

Sayers, R. R., W. P. Yant, B. G. H. Thomas, and L. B. Berger: Physiological response attending exposure to vapors of methyl bromide, methyl chloride, ethyl bromide and ethyl chloride. Public Health Bull. **185**, 1-8, 20-40 (1929).

Schifferli, E.: Intoxications par le bromure de methyle. Rev. Med. Suisse Romande **62**, 244 (1942).

Schmittle, S. C.: Studies on methyl bromide. I. The efficacy of methyl bromide fumigation on Newcastle disease virus. Poultry Sci. **34**, 1219 (1955).

Schuler: Vergiftung durch Brommethyl? Deutsche Vierteljahrsschrift fuer Offentliche Gesundheitspflege **31**, 696 (1899).

Scudamore, K. A., and S. G. Heuser: Residual free methyl bromide in fumigated commodities. Pest. Sci. **1**, 14 (1977).

Shield, L. K., T. L. Coleman, and W. R. Markesbery: Methyl bromide intoxication: neurologic features, including simulation of Reye syndrome. Neurology **27**(10), 959 (1977).

Shrader, S. A., A. W. Beshgetoor, and V. A. Stenger: Determination of total and inorganic bromide in foods fumigated with methyl bromide. Anal. Chem. **14**, 1 (1942).

Sigwald, J., D. Bouttier, J. Guibert, and P. J. Nicolas-Charles: Intoxication aigue par le bromure de methyle. Amelioration rapide des sequelles par des fortes doses de pyridoxine. Rev. Neurol. **92**, 270 (1955).

Simmon, V. F., and R. Tardiff: The mutagenic activity of halogenated compounds found in chlorinated drinking water. In R. Jolley (ed.): Water chlorination: Environmental impact and health effects, Vol. 2, p. 417. Ann Arbor: Ann Arbor Science Publishers (1978).

_______, K. Kauhanen, and R. G. Tardiff: Mutagenic activity of chemicals identified in drinking water. Dev. Toxicol. Environ. Sci. 2 (Prog. Genet. Toxicol.), 249 (1977).

Skimina, C.: Weed control in container crops. Proc. Ann. Calif. Weed Conf. **29**, 137 (1977).

Snitko, V. M., and Y. E. Leuchenko: Fumigant reducing the hazard of disinfection operations and residual bromides in products. Gig. Sanit. **38**(12), 96 (1973).

Soremark, R.: The biological half-life of bromide ions in human blood. Acta Physiol. Scand. **50**, 119 (1960 a).

_______ Distribution and kinetics of bromide ions in the mammalian body. Some experimental investigations using Br 80M and Br 82. Acta Radiol. Suppl. **190**, 1 (1960 b).

Spencer, H. C., V. K. Rowe, E. M. Adams, and D. D. Irish: Response of rats maintained on diets fumigated with methyl bromide. Food Res. **9**, 11 (1944).

Starratt, A. N., and E. J. Bond: Metabolism of methyl bromide by susceptible and resistant strains of the granary weevil, *Sitophilus granarius* (L.). Pest. Biochem. Physiol. **15**(3), 275 (1981).

Stenger, V. A.: Bromine compounds. In Kirk-Othmer: Encyclopedia of chemical technology, 3rd ed., Vol. 4, pp. 249-263. New York: Wiley (1978).

Takahashi, S., T. Moroji, and T. Yamauchi: A case of methyl bromide intoxication which showed paroxysmal abnormal EEG. Rinsho Noha. **15**(11), 725 (1973) [cited in Pest. Abstr. **74**, 1614 (1973)].

Takayama, J., and S. Numajiri: A serious case of methyl bromide intoxication. Nippon Shonika Gakkai Zasshi. **83**(3), 332 (1979) [cited in Pest. Abstr. **79**, 1595 (1979)].

Thiebaut, F., F. Daum, and H. Henrot: Tremblement spasmodique intentionnel des membres superieurs, consecutif a une intoxication par le bromure de methyle. Exemple de syndrome strie d'origine toxique, rappelant le tremblement de la pseudoscherose de Westphal-Strumpell. Rev. Neurol. **75**, 35 (1943).

Thomson, G. R.: A case of burning and slight poisoning by methyl bromide. Brit. J. Surg. **33**, 91 (1945).

Thompson, R. H.: A review of the properties and usage of methyl bromide as a fumigant. J. Stored Prod. Res. **1**, 353 (1966).

Tietze, A.: Klinische Beobachtungen zur Methylbromid und Tetrachlorkohlenstoff-vergiftung. Arch. Gewerbepathol. Gewerbehyg. **4**, 733 (1933).

Toennies, G., and J. J. Kolb: Methionine studies. VII. Sulfonium derivatives. J. Amer. Chem. Soc. **67**, 849 (1945).

Toyonaga, K., and S. Tokuda: Physiological examination and treatment by clonazepam of action myoclonus due to methyl bromide intoxication. Rinsho Shinkeigaku. **16**(11), 830 (1976) [cited in Pest. Abstr. **77**, **1034** (1976)].

Uchiyama, H., M. Sato, H. Momotani, S. Morinobu, and S. Ishizu: On the amount of bromide in the blood of animals exposed to aerial methyl bromide. Sangyo Igaku. **21**(4), 662 (1979) [cited in Pest. Abstr. **80**, 8049 (1979)].

Ucko, H.: Investigations into the presence and role of bromine in the body. Biochem. J. **30**, 992 (1936).

United States Department of Agriculture, Agriculture Stabilization and Conservation Service: The pesticide review, p. 41. Washington (1970).

______ The pesticide review, p. 32. Washington (1978).

United States Department of Health, Education and Welfare, Public Health Service, Center for Disease Control: Morbidity and mortality report. **23**(25), 217 (1974).

United States Environmental Protection Agency, Office of Pesticide Programs: The EPA compendium of registered pesticides, Vols. II, III, IV, V, VI. Washington: U.S. Government Printing Office (1974).

Ushio, K., and R. Osozuka: A case of severe intoxication due to methyl bromide. Sangyo Igaku. **19**(5), 355 (1977) [cited in Pest. Abstr. **78**, 1083 (1977)].

Van Assche, C.: Behavior and perspectives of chemical soil fumigation. Proc. 6th Brit. Insect. Fumg. Conf., pp. 706-714 (1971).

Van Gundy, S. D., and R. L. Van Gundy: Methyl bromide soil fumigation for continuous Christmas tree (*Pinus radiata*) production. Down to Earth **27**(2), 1 (1971).

Van Logten, M. J., M. Wolthus, A. G. Rauws, and R. Kroes: Short-term toxicity study in rats. Toxicology **1**, 321 (1973).

Verberk, M. M., T. Rooyakkers-Beemster, M. De Vlieger, and A. G. M. Van Vliet: Bromine in blood, EEG and transaminases in methyl bromide workers. Brit. J. Ind. Med. **36**(1), 59 (1979).

Viner, N.: Methyl bromide poisoning: A new industrial hazard. Can. Med. Assoc. J. **53**, 43 (1945).

von Oettingen, W. F.: The toxicity and potential dangers of methyl bromide with special reference to its use in the chemical industry in fire extinguishers and in fumigation. U.S. National Institute of Health Bull. No. 185, pp, 1-41 (1946).

______ The halogenated aliphatic, olefinic, cyclic, aromatic, and aliphatic-aromatic hydrocarbons including the halogenated insecticides, their toxicity and potential dangers. U.S. Public Health Service Publ. 414, pp. 15-30 (1955).

______ Poisoning: A guide to clinical diagnosis and treatment, pp. 429-431. Philadelphia: Saunders (1958).

Watrous, R. M.: Methyl bromide: Local and mild systemic toxic effects. Ind. Med. **11**(12), 575 (1942).

Weast, R. C. (ed.): Handbook of chemistry and physics. 62nd Ed., p. C-371. Cleveland: Chemical Rubber (1981-1982).

White, J. G.: Weed control by soil partial sterilant chemicals: A review. Proc. Brit. Crop Prot. Conf. Weeds **14**(3), 987 (1978).

Wikoff, H. L., R. A. Brunner, and H. W. Allison: The normal bromine content of the blood of healthy individuals. Amer. J. Clin. Pathol. **10**, 234 (1940).

Williams, N., and W. Reilly: Methyl bromide poisoning. Amer. Ind. Hyg. Assoc. J. **31**(2), 30 (1970).

Williford, J. H. Jr., R. F. Kelly, and P. P. Graham: Residual bromide in tissues of rats fed methyl bromide fumigated diets. J. Animal Sci. **38**(3), 572 (1974).

Winter, R. M.: The toxicity of methyl bromide. The Chem. Trade J. and Chem. Eng. **98**, 471 (1936) [cited in von Oettingen (1946)].

Winteringham, F. P. W.: The fate of labelled insecticide residues in food products. IV. The possible toxicological and nutritional significance of fumigating wheat with methyl bromide. J. Sci. Food Agr. **6**, 269 (1955).

______, and J. M. Barnes: Comparative response of insects and mammals to certain halogenated hydrocarbons used as insecticides. Physiol. Rev. **35**, 701 (1955).

______, and A. Harrison: The sorption of methyl bromide by wheat. J. Soc. Chem. Ind. **65**, 140 (1946).

______ ______, R. G. Bridges, and P. Bridges: The fate of labelled insecticide residues in food products. II. The nature of methyl bromide in residues in fumigated wheat. J. Sci. Food Agr. **6**, 251 (1955).

Wyers, H.: Methyl bromide intoxication. Brit. J. Ind. Med. **2**, 24 (1945).

Yomura, Y., and N. Shibata: Studies on the intoxication due to the intake of foods fumigated with methyl bromide. Part 1. A case of intoxication due to methyl bromide, which started the studies. St. Marianna Ikadaigaku Zasshi. **6**(3), 389 (1978) [cited Pest. Abstr. **79**, 859 (1978)].

______ ______, and E. Toguri: On the toxicity of food fumigated with methyl bromide. Jpn. J. Med. **15**(3), 281 (1976).

Yoshida, Y., K. Kono, and A. Harada: Cases of suspected methyl bromide poisoning in harbor workers and a survey of their occupational environment. Sangyo Igaku **17**(1), 3 (1975) [cited in Pest. Abstr. **75**, 2143 (1975)].

Zatuchni, J., and K. Hong: Methyl bromide poisoning seen initially as psychosis. Arch. Neurol. **38**(8), 529 (1981).

Zondek, H., and A. Bier: Brom in Blute bei Manisch-depressiuem irresein. Biochem. Z. **241**, 491 (1931).

Manuscript received November 4, 1982; accepted November 15, 1982

Subject Index

INFORMATION FOR AUTHORS

RESIDUE REVIEWS

(A BOOK SERIES CONCERNED WITH RESIDUES OF PESTICIDES AND OTHER CONTAMINANTS IN THE TOTAL ENVIRONMENT)

Edited by

Francis A. Gunther
Jane Davies Gunther

Published by
Springer-Verlag New York • Heidelberg • Berlin

The original (ribbon) copy and one good xerox or other copy of the manuscript, complete with figures and tables, are required. Manuscripts will normally be published in the order in which they are received, reviewed, and accepted. They should be sent to the editor:

Professor Francis A. Gunther
Department of Entomology
University of California
Riverside, California 92521
Telephone: (714) 787-5804/5810 (office)
(714) 688-6666 (home)

1. Manuscript

The manuscript, in English, should be typewritten, double-spaced throughout, on one side of 8½ x 11-inch blank white paper, with at least one-inch margins. The first page of the manuscript should start with the title of the manuscript, name(s) of author(s), with author affiliation(s) as first-page starred footnotes, and "Contents" section. Pages should be numbered consecutively in arabic numerals, including those bearing figures and tables only. In titles, in-text outline headings and subheadings, figure legends, and table headings only the initial word, proper names, and universally capitalized words should be capitalized.

Footnotes should be inserted in text and numbered consecutively in the text using arabic numerals.

Tables should be typed on separate sheets and numbered consecutively within the text in *roman numerals*; they should bear a descriptive heading, in lower case, which is underscored with one line and which starts after the word "Table" and the appropriate roman numeral; *footnotes in tables* should be designated consecutively within a table by the lower-case alphabet. *Figures* (including photographs, graphs, and line drawings) should be numbered consecutively within the text in arabic numerals; each figure should be affixed to a separate page bearing a legend (below the figure) in lower case starting with the term "Fig." and a number.

2. Summary

A concise but informative summary (double-spaced) must conclude the text of each manuscript; it should summarize the significant content and major conclusions presented. It must not be longer than two 8½ × 11-inch pages of double-spaced typing. As a summary, it should be more informative than the usual abstract.

3. References

All papers, books, and other works cited in the text must be included in a "References" section (*also double-spaced*) at the end of the manuscript: If comprehensive papers on the same subject have been published, they should be cited but only for exceptional reasons should the bibliographic citations extend farther back than to these papers.

The references used *in the text* should consist of the author's or authors' last name(s) where one or two authors are concerned; should there be more than two authors, only the first is named and "*et al.*" is added. The publication year in parentheses should follow the name. If more than one paper by one author published in the same year is cited, the letters a, b, c, etc., should follow the year, e.g., "Meier (1958 a) found . . .", or "This method is nonspecific (Meier 1958 a)."

In the References section, the papers cited should appear in alphabetical order according to the last name of the first author; if more than one paper by an author or authors published in the same year is cited, the papers should be listed according to the year of publication followed by a, b, c, etc., as necessary. Papers published in periodicals should be cited with names and initials of all authors, together with the *full title of the paper and preferably in its original language,* title of the periodical (abbreviated in accordance with *Chemical Abstracts'* "List of Periodicals Abstracted"), number of the volume (wavy underlined), initial page, and the year in parentheses. References to unpublished papers that have been submitted for publication should be cited in the same manner as other papers except the abbreviated journal name is followed by the words "In press" or "Accepted for publication" and the year in parentheses; personal communications are to be cited similarly.

In text and in the References section, citation of *governmental agencies, educational and research institutions and foundations, professional associations, and industrial companies* should consist of the full name as used by the organization, completely underscored with one line and with initial capital letters only, followed by the appropriate reference information as specified above.

Examples:

Edwards, C. A., and E. B. Dennis: Some effects of aldrin and DDT on the soil fauna of arable land. Nature 188, 767 (1960).

Gunther, F. A., J. H. Barkley, and W. E. Westlake: Worker environment research. II. Sampling and processing techniques for determining dislodgable pesticide residues on leaf surfaces. Bull. Environ. Contam. Toxicol. Accepted for publication (1974).

Hessler, W.: Eine einfache Nachweismethode für Paraffin in Wachsgemischen. II. Mitt. Fette, Seifen, Anstrichmittel 58, 602 (1956).

Melzer, H.: The qualitative and quantitative colorimetric determination of captan. Nachrbl. deut. Pflanzenschutzdienst 14, 193 (1960).

Shell Chemical Co.: Letter to EPA's "Hazardous Materials Advisory Committee," Oct. 28 (1971).

U.S. Environmental Protection Agency: Proposed toxicology guidelines. Fed. Register 37 (183), 19383 (1972).

Books should be cited with name(s) and initials of the author(s), full title, edition or volume, page number(s), place of publication, publisher, and year of publication in parentheses.

Examples:

Bevenue, A.: Gas chromatography. In G. Zweig (ed.): Analytical methods for pesticides, plant growth regulators, and food additives. Vol. I, p. 189. New York: Academic Press (1963).

Dormal, S., and G. Thomas: Répertoire toxicologique des pesticides, p. 48. Gembloux: J. Duculot (1960).

Harte, C.: Physiologie der Organbildung, Genetik der Samenpflanzen. In: Fortschritte der Botanik. Vol. 22, p. 315. Berlin-Göttingen-Heidelberg: Springer (1960).

Metcalf, R. L.: Organic insecticides, their chemistry and mode of action. 2 ed., p. 51. New York-London: Interscience (1961).

4. Illustrations

Illustrations of any kind may be included only when indispensable for the comprehension of text; they should not be used in place of concise, clear explanations in text. Schematic line drawings must be drawn carefully and clearly. For other illustrations, clearly defined black-and-white glossy photographic prints are required. Should precisely placed indication darts (arrows) or letters be required on a photograph or other type of illustration, they should be marked neatly with a soft pencil on a duplicate copy or on an overlay, with the end of each dart (arrow) indicated by a fine pinprick; darts and lettering will be transferred to the illustrations by the publisher.

Photographs should be not less than five × seven inches in size. Unimportant and indistinct strips or areas on the edges of photographs should be marked on the back of the glossy print (pattern) with pencilled down-strokes, in order that the reproduction surface will not be unnecessarily large; alterations of photographs in page proof stage are not permitted. *Each photograph or other illustration should be marked on the back, distinctly but lightly, with soft pencil, with first author's name, figure number, manuscript page number, and the side which is the top.*

If illustrations from published books or periodicals are used, the exact source of each should be included in the figure legend; if these "borrowed" illustrations are copyrighted by others, permission of the copyright holder to reproduce the illustration must be secured by the author.

5. Nomenclature

All pesticides and other subject-matter chemicals should be identified according to *Chemical Abstracts,* with the full chemical name in text in parentheses or brackets the first time a common or trade name is used. *If many such names are used, a table of the names and their precise chemical designations should be included as the last table in the manuscript, with a numbered footnote reference to this fact on the first text page of the manuscript.*

6. Miscellaneous

Abbreviations. Common units of measurement and other commonly abbreviated terms and designations should be abbreviated as listed below; if any others are used often in a manuscript, they should be written out the first time used, followed by the normal and acceptable abbreviation in parentheses [e.g., Acceptable Daily Intake (ADI), Angstrom (Å), picogram (pg), parts per trillion (ppt)]. Except for inch (in.) and number (no., when followed by a numeral), abbreviations are used without periods. Temperatures should be reported as "°C" or "°F" (*e.g.,* mp 41° to 43°C).

Abbreviations

A	acre	μg	microgram(s)
bp	boiling point	μL	microliter(s)
cal	calorie	μm	micrometer(s)
cm	centimeter(s)	mg	milligram(s)
cu	cubic (as in "cu m")	mL	milliliter(s)
ft	foot (feet)	mm	millimeter(s)
gal	gallon(s)	m<u>M</u>	millimolar
g	gram(s)	min	minute(s)
ha	hectare	<u>M</u>	molar
hr	hour(s)	mon	month(s)
in.	inch(es)	ng	nanogram(s)
id	inside diameter	nm	nanometer(s) (millimicron)
kg	kilogram(s)	<u>N</u>	normal
L	liter(s)	no.	number(s)
mp	melting point	od	outside diameter
m	meter(s)	oz	ounce(s)

Abbreviations

ppb	parts per billion
ppm	parts per million
/	per
lb	pound(s)
psi	pounds per square inch
rpm	revolutions per minute
sec	second(s)
sp gr	specific gravity
sq	square (as in "sq m")
vs.	versus
wk	week(s)
wt	weight
yr	year(s)

Numbers. All numbers used with abbreviations and fractions or decimals are arabic numerals. Table numbers are roman numerals. Otherwise, numbers below ten are to be written out. Numerals should be used for a series (e.g., "0.5, 1, 5, 10, and 20 days"), for pH values, and for temperatures. When a sentence begins with a number, write it out.

Symbols. Special symbols (*e.g.,* Greek letters) must be identified in the margin, *e.g.,*

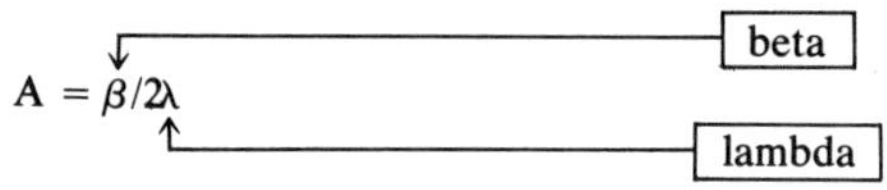

Percent should be % in text, figures, and tables.

Style and format. The following examples illustrate the style and format to be followed (except for abandonment of periods with abbreviations):

Kaemmerer, K., and S. Buntenkötter: The problem of residues in meat of edible domestic animals after application or intake of organophosphate esters. Residue Reviews 46, 1 (1973).

The Chemagro Division Research Staff: Guthion (azinphosmethyl): Organophosphorus insecticide. Residue Reviews 51, 123 (1974).

7. Proofreading scheme

The senior author must return the Master set of page proof to the Editor within one week of receipt. Author corrections should be clearly indicated on proof with soft pencil or with ink, and in conformity with the standard "Proofreader's Marks" accompanying each set of proofs. In correcting proof, new or changed words or phrases should be carefully and legibly handprinted (not handwritten) in the margins.

8. Offprints

Senior authors receive 30 complimentary offprints of a published article. Additional offprints may be ordered from the publisher at the time the principal author receives the proof. Order forms for additional offprints will be sent to the senior author along with the page proofs.

9. Page charges

There are no page charges, regardless of length of manuscript. However, the cost of alterations (other than corrections of typesetting errors) attributable to authors' changes in the page proof, in excess of 10% of the original composition cost, will be charged to the authors.

If there are questions that are not answered in this leaflet, see any volume of *Residue Reviews* or telephone the Editor (see p. 1 for telephone numbers). Volume 3 (Ebeling) is especially helpful for style and format.